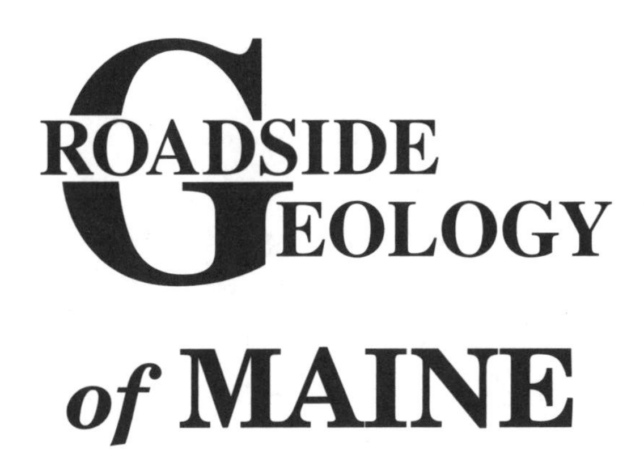

D. W. Caldwell

Mountain Press Publi
Missoula, Montana
1998

Sixth Printing, April 2014

Cover illustration and design by Kim Ericsson

Library of Congress Cataloging-in-Publication Data

Caldwell, Dabney W., 1927–2006
Roadside geology of Maine / D.W. Caldwell.
p. cm.
Includes bibliographical references and index.
ISBN 978-0-87842-375-0 (alk. paper)
1. Geology—Maine—Guidebooks. 2. Maine—Guidebooks.
I. Title.
QE119.C35 1998 98-17184
557.41—dc21 CIP

PRINTED IN THE UNITED STATES

P.O. Box 2399 • Missoula, MT 59806 • 406-728-1900
800-234-5308 • info@mtnpress.com
www.mountain-press.com

This book is dedicated to the teachers who made it possible:

Schools in Mount Vernon, Maine
Hazel Cole
Mildrid Scarci
Barbara Williams

Kents Hill School, Kents Hill, Maine
Pug Goldthwait

Bowdoin College, Brunswick, Maine
Herby Brown
R. P. T. Coffin

Brown University, Providence, Rhode Island
Lon Quinn
Phil Shafer

Harvard University, Cambridge, Massachusetts
Marland Billings
Johnny Miller

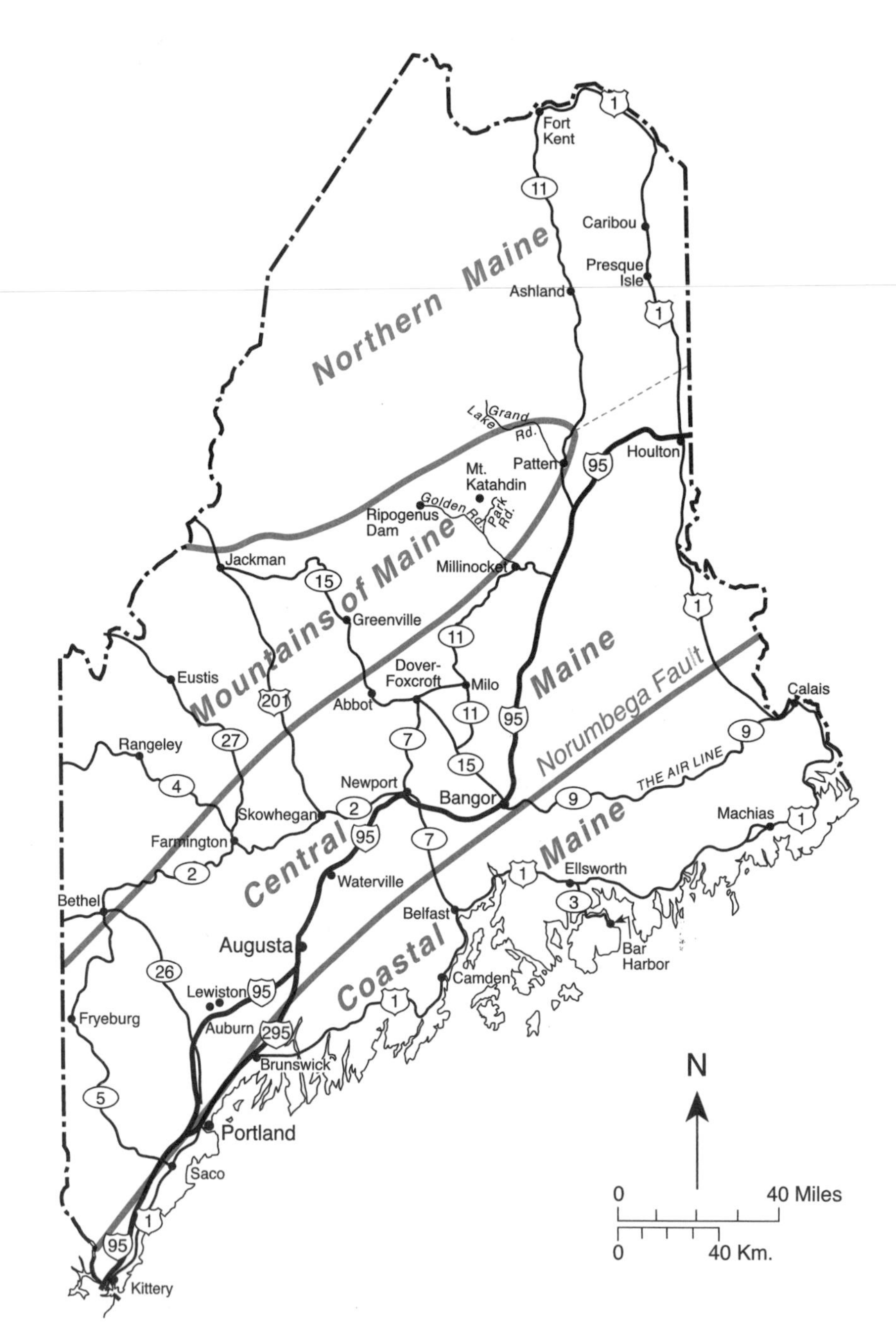

Roads and sections of Roadside Geology of Maine.

Contents

Preface

In planning this book I tried to cover every major geologic feature of Maine, but I could not describe the rocks along every road in the state. It was also not my intent to cover the complete length of even the major roads. Only the 300-mile length of Interstate 95 in Maine is described completely. The geology along nearly 400 miles of U.S. 1 in Maine is covered, missing only the section between Calais and Houlton. On most roads the rocks are described from south to north. U.S. 2 between Skowhegan and the New Hampshire border is discussed from east to west. The book follows the Air Line, Maine 9, from Bangor to Calais from west to east. In the northern Maine wilderness, I discuss the geologic features along some dirt roads in and around Baxter State Park.

Some of the roads mentioned are a bit off the beaten track, poorly marked, and do not appear on ordinary road maps. I strongly recommend DeLorme's *The Maine Atlas and Gazetteer* or a set of U.S. Geological Survey topographic maps as an indispensable accessory. Perham's Mineral Store at Trap Corner in West Paris is a good source of information about all of the pegmatites in Maine. They own several pegmatite quarries, in which collecting, identification of specimens, and other advice are free.

Always wear safety glasses or goggles when hammering on rocks. Flying splinters of rock or steel can severely damage your eyes.

Professor Lindley Hanson contributed greatly to the early phases of the planning and writing of this book. Unfortunately, other commitments made it impossible for her to continue as coauthor. George and Maggie Hanson of Acworth, New Hampshire, provided a quiet sanctuary where most of the writing was done. Sal Rand of Freeport gave welcome advice on the history of Maine. She also lent me her copy of *Maine, A Guide Downeast,* a book written during the Depression by members of the Writers' Project of the Work Projects Administration. This marvelous work was reprinted and updated in 1972. Jack Rand of Freeport, former state geologist of Maine, pointed out a number of areas in which my understanding of the state's geological features was less than satisfactory. The Maine Geological Survey—Walter Anderson, state geologist—recently published two geologic maps of the state and an impressive six-volume work, *Studies in Maine Geology,* all of which I used extensively. The New England Intercollegiate Geological Conference has conducted field trips in Maine six times since 1980. The guidebooks from those meetings provide thorough information on the geology of all parts of Maine.

Eliza McClennen did the artwork on the geologic maps and most of the line drawings.

Era	Period	Events in Maine	Time (years ago)
CENOZOIC	Quaternary	Ocean rises to present level. Rivers take on modern courses. Glaciers disappear from Maine. Last glaciers appear in Maine.	2,000 11,000 12,000 30,000 ?
	Tertiary	No record in Maine.	
		Cenozoic–Mesozoic Boundary	**66 mya***
MESOZOIC	Cretaceous	Igneous rocks in York County.	68 to 100 mya
	Jurassic	Basalt dikes near coast accompany rifting of Pangea.	190-200 mya
	Triassic	Agamenticus igneous complex.	228 mya
		Mesozoic–Paleozoic Boundary	**245 mya**
PALEOZOIC	Permian	Pegmatites in Topsham.	298 mya
	Carboniferous	Formation of Sebago granite.	325 mya
	Devonian	Acadian orogeny: Folding and faulting of all rocks older than early Devonian. Formation of most granites in Maine.	365 to 405 mya
	Silurian	Deposition of sandstone and shale, with local ribbon rock in central and northern Maine. Volcanism in east coastal Maine.	425 mya
	Ordovician	Taconic orogeny: Volcanic rocks in the Lobster Mtn.–Winterville belt. Volcanism and deposition of Casco Bay formation.	450 mya
	Cambrian	Deposition of Grand Pitch, Hurricane Mtn., and Jim Pond formations in mountains of Maine. Volcanism and deposition of Ellsworth formation.	505 mya
		Paleozoic–Precambrian Boundary	**542 mya**
PRECAMBRIAN		Pegmatite intrudes Seven Hundred Acre Island formation, Penobscot Bay.	647 mya
		Grenville orogeny and formation of basement rocks under northern Maine.	1,000 mya
			*million years ago

Geologic time scale.

Geology of Maine

Six hundred million years ago, an ocean lay where Maine and New England are now. The rocks supporting the New England landscape existed then as island archipelagos and as mud on the ocean floor. The eastern margin of our ancient continent lay near Québec City and Albany. In less than 400 million years the ocean closed in a series of events that piled sediment and island fragments against the ancient margin of the continent. These events also melted rock, so that volcanic rocks were added, and granite—plenty of granite.

No coast was nearby when the old ocean finally closed, and the assembly of Maine was mostly finished by about 250 million years ago. West Africa was adjacent to New England. Most of the earth's continental crust was assembled into the supercontinent, Pangea. Nova Scotia was attached to Europe. Florida was attached to both Africa and South America. Asia was attached to Europe, as it still is. And India, Antarctica, and Australia were attached to the eastern side of Africa. Maine was closer to the equator. The earliest dinosaurs must have roamed then, but no traces of them have been found in Maine.

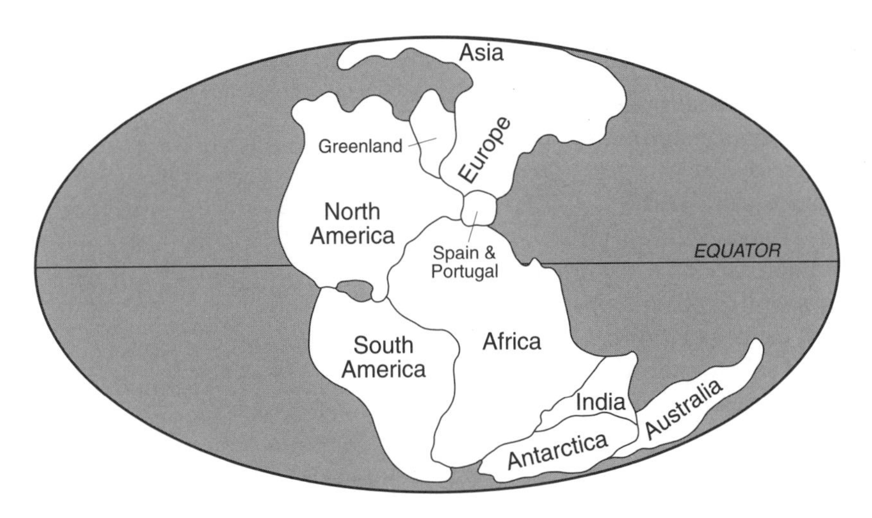

Pangea about 225 million years ago.

Pangea began to break up about 200 million years ago. Rifts opened between North America, Europe, and Africa, and the Atlantic Ocean began to open between them, its floor forming through volcanic eruptions at the rifts. North America and Europe moved northwest while Africa moved southeast. They are still moving, and the Atlantic Ocean is still growing.

The rocks of Maine record these events, as well as the ice ages of the last 2 million years. Maine is an intricate collage of fragments, each with its own complex past. Geologists argue about those records as they try to decipher them. Many mysteries will remain, because too many pieces of the puzzle are missing, or simply refuse to lie flat. The geologic history presented here is controversial in places, but most geologic ideas are, especially after they appear in print.

The Iapetus Ocean

The story of how the pieces of Maine came together begins about 600 million years ago, when an ancient supercontinent, Rodinia, broke apart. *Rodinia* is Russian for "motherland." It split to the accompaniment of great volcanic eruptions that produced immense basalt lava flows. Some of the molten basalt crystallized within fractures in the Grenville basement rocks to become great dikes.

Then the pieces of the Rodinia continent drifted apart as basalt lava flows continued to erupt from the rift between them, making the floor of the widening Iapetus Ocean, which grew for millions of years. No one knows how wide it became. It was a complex ocean with marginal seas and several islands that were in fact fragments of continental crust.

The Iapetus Ocean was mainly in the Southern Hemisphere, with large continents north and south. Laurentia, predecessor to North America, lay near the equator, bordering the ocean's northern margin. Across the ocean farther south lay two large continents and a microcontinent. The two large continents were Gondwana and Baltica, predecessors of Africa and Northern Europe. Avalonia is the name given to the microcontinent. It would eventually collide with Laurentia and later be carried off as part of Europe.

The Iapetus Ocean grew to be so broad that organisms could not cross it. Faunas on its opposite sides followed their own evolutionary paths. Animals associated with the southern continents, the Acado-Baltic fauna, developed quite differently from those in the waters off ancestral North America. These fossils were entombed in rocks that were far apart and then brought together during the continental collisions of middle Paleozoic time. The juxtaposition of rocks containing fossils of these distinct faunas was the first major clue that led geologists to the discovery that the Appalachians were created in the destruction of an ocean.

Although geologists call it Laurentia, the continent that lay on our side of the Iapetus Ocean was, for all practical purposes, North America. Even so, its size and shape were far different from that of the continent we now know. The rocks of Maine are later additions to its eastern margin, which is now in southeastern Québec. During early Paleozoic time, perhaps 500 million years ago, the Iapetus Ocean was growing and this margin faced southeast, along the equator. It was deeply embayed, with great promontories and reentrants. These large headlands and embayments would play an important role in determining the intensity of mountain building and the timing of deformation while the ocean closed.

Iapetus Ocean Begins to Close

Oceans open and oceans close. As one ocean grows wider, so must another grow narrower. New ocean floor grows where basalt flows erupt from between spreading plates. Meanwhile, old ocean floor sinks

Rifting and seafloor spreading (top) *and subduction and seafloor destruction* (bottom).

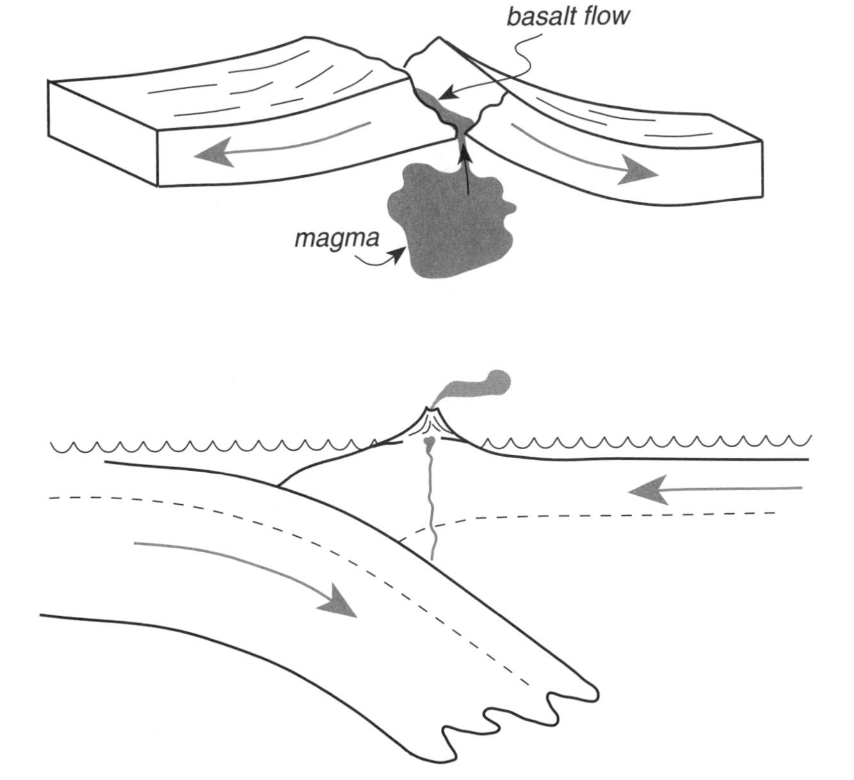

through oceanic trenches and plunges into the mantle. The old oceanic lithosphere absorbs heat as it sinks into the mantle: it finally disappears when it reaches the temperature of the surrounding rocks.

On the other side of the Iapetus Ocean was Avalonia, among other and much larger continents. It would eventually collide with the continental margin of Laurentia as the last of the floor of the Iapetus Ocean that had separated them sank through an oceanic trench. Sometime late in Cambrian time, perhaps about 520 million years ago, the Iapetus Ocean began to close as its floor sank along oceanic trenches on its opposite sides. The oldest evidence of that sinking is volcanic rocks, the kind that erupt from a volcanic chain above a sinking oceanic plate. Chains of volcanoes developed that must have resembled the modern Aleutian chain, which stands above the sinking ocean floor and aligns parallel to the Aleutian trench. The Iapetus Ocean closed not all at once, but in a piecemeal fashion.

Penobscot Mountain-Building Event

Meanwhile, somewhere in the Iapetus Ocean and sometime during late Cambrian or earliest Ordovician time, two miniature continents collided and welded together. We know them now as the Boundary Mountains and Gander terranes. The collision that joined them is the Penobscot mountain-building event. A belt of crushed and cleaved oceanic rock and marine sediment called the Hurricane Mountain mélange defines the suture between the Boundary Mountains and Gander terranes. It is essentially a last bit of old ocean floor caught in the vice between the colliding continental fragments.

We know that this collision occurred far from Laurentia because soon after the mountain-building event, the newly consolidated continent was partially submerged and covered with muddy sediments. Unlike the sediments that were then accumulating along the margin of Laurentia, these contain the fossils of animals that probably lived in cooler water, presumably at some higher latitude. The composite Boundary Mountains and Gander block was destined to get caught between the colliding continents of the Taconic mountain-building event.

Taconic Mountain-Building Event

The Taconic mountain-building event happened during middle Ordovician time, between about 480 and 430 million years ago. The seafloor of the Iapetus Ocean slid through an oceanic trench and sank beneath the composite block of the Boundary Mountains and Gander terranes. A chain of volcanoes developed above the sinking oceanic crust, on top of the overriding Boundary Mountains and Gander block. And as the oceanic trench gobbled the seafloor that separated them, the Boundary Mountains and Gander block slowly approached Laurentia, which was rapidly assuming its new identity as ancient North America.

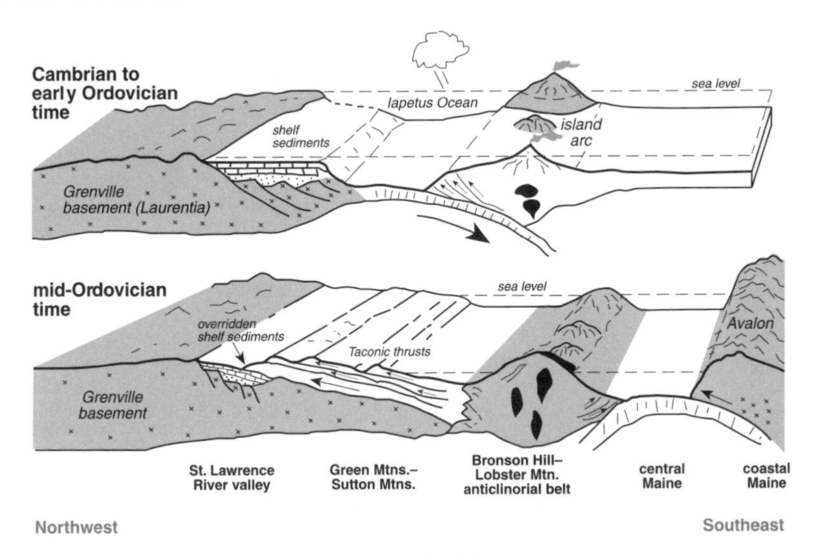

Taconic mountain-building event.

As the last of the intervening ocean floor sank through the trench, the Boundary Mountains and Gander block jammed into the edge of North America. The collision caused the continental margin to sink and the remaining seafloor to buckle up between North America and the Boundary Mountain and Gander block. A flood of muddy sediments that had eroded from the highlands and that rose along the line of collision flowed northwest to bury the limestone reefs that had once thrived along the coast. Broken slabs of oceanic crust were thrust over the muddy sediments, over the continental shelf, and onto North America. A narrow belt of stranded oceanic crust, called ophiolite, traces the suture where the continental blocks joined. This tectonic seam appears in central Vermont, in southern Québec, and in a small area of northwestern Maine; it is known as the Baie Verte–Brompton Line.

Rocks of the Boundary Mountains and Gander block are now in western and central Maine, southeast of their suture with the old North American continent. They include the basement rocks of the Chain

Lakes massif and the Hurricane Mountain mélange. And the same area contains Ordovician plutonic and volcanic rocks that erupted about 480 million years ago, the eroded roots of a volcanic chain. Together, these rocks compose the core of the central Maine volcanic province, which trends northeast through Maine from the New Hampshire border. The trend of these Ordovician igneous rocks in New England clearly shows the position of the old volcanic chain. This province would become the site of more volcanic activity during the Acadian mountain-building event.

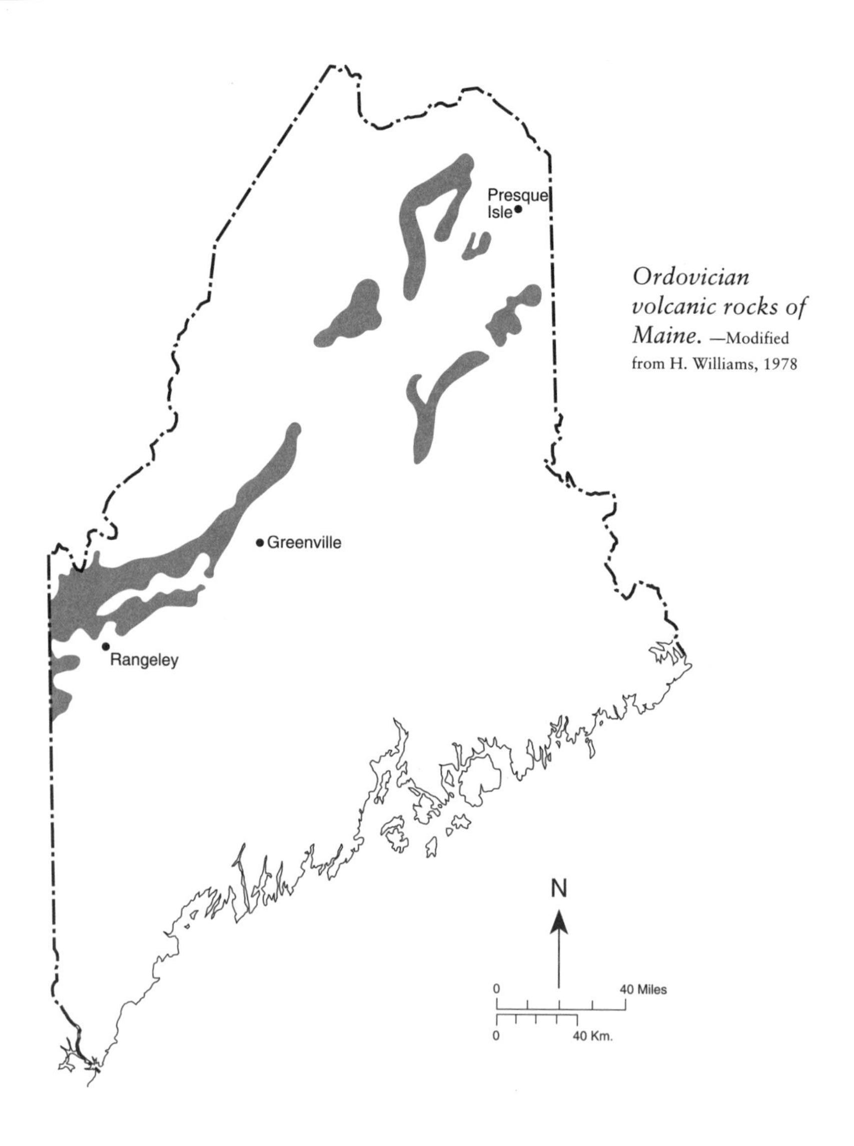

Ordovician volcanic rocks of Maine. —Modified from H. Williams, 1978

Many geologists view the Taconic mountain-building event somewhat differently, as essentially the first stage in the collision of North America and Europe. Some contend that the Iapetus Ocean closed but that the collision was incomplete. It happened as the projecting promontories of the two continents ground into each other, leaving deep basins of ocean floor in the bays between them. Those basins continued to fill with sediment that was finally crushed in the Acadian continental collision. Some geologists contend that the Taconic mountain-building event was caused by the closing of a smaller marginal sea within the Iapetus Ocean and that the principal ocean still remained, albeit smaller. And still others believe that rifting of the newly sutured continent created the ocean floor that the Acadian mountain-building event would destroy. At any rate, most will agree that the

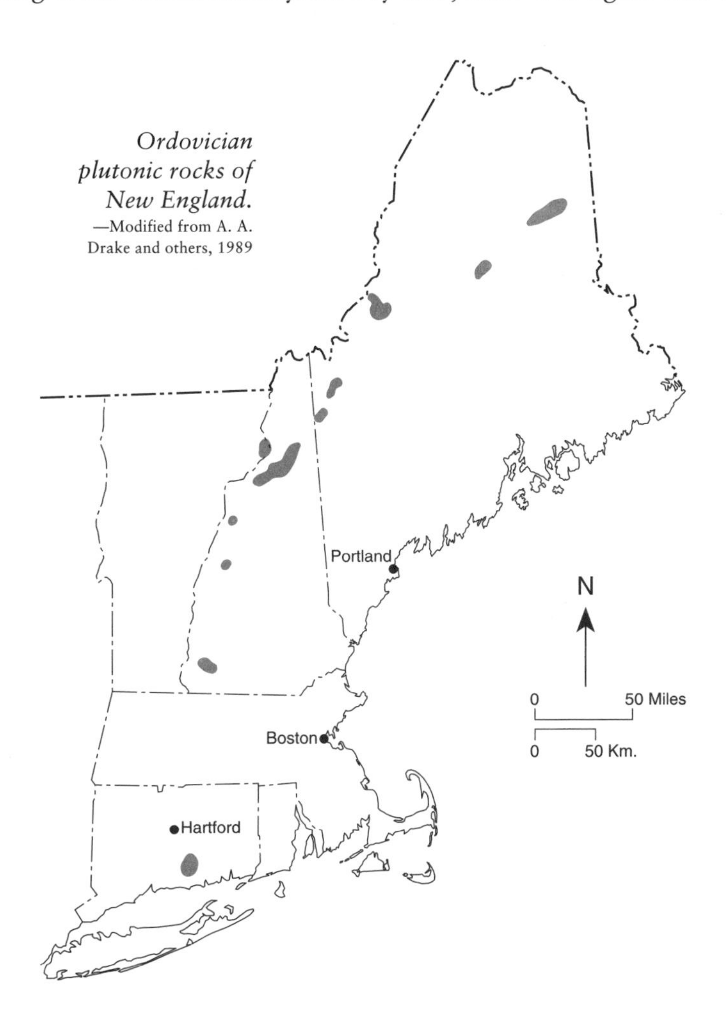

Ordovician plutonic rocks of New England. —Modified from A. A. Drake and others, 1989

final and complete closure that added the last pieces of Maine happened long after the Taconic mountain-building event.

Silurian Time

Immediately after the Taconic mountain-building event, sometime in late Ordovician and early Silurian time, perhaps about 425 million years ago, several deep basins formed and began filling with sediment. During early Silurian time, about 440 to 420 million years ago, the newly raised Taconic mountains shed sediment eastward. It accumulated in great submarine fans along the new margin of North America and spread east across the floor of the Iapetus Ocean—what remained of it.

Sometime in middle Silurian time, about 420 million years ago, another large basin called the Connecticut Valley–Gaspé trough started growing. A rift with new ocean floor forming within it separated the volcanic chain of Ordovician time onto opposite shores of the new ocean. Muddy and sandy sediments were dumped on the basalt seafloor. That continued through late Silurian and early Devonian time, until the Acadian mountain-building event crushed those sediments into tightly folded sandstones and slates. You can see those rocks now in the Connecticut Valley–Gaspé synclinorium, a major structural feature of the Northern Maine slate belt.

While basins were forming on the continent, one or more oceanic trenches were swallowing the last of the floor of the Iapetus Ocean, pulling closer the continents that lay across the narrowing ocean. Chains of volcanoes grew parallel to those trenches, one on the margin of the approaching Avalon continent, the other along the trend of the central volcanic province, the old volcanic chain of Taconic time. As late Silurian time began, an influx of sediment washing in from an eastern source heralded the arrival of the approaching Avalon continent. By the close of late Silurian time, the remaining floor of the Iapetus Ocean plunged through an oceanic trench and into the hot depths of the earth's mantle.

The Avalon terrane depressed the edge of North America as it collided. The continent sank through all of what is now central and northeastern Maine. Great volumes of sediment turbidity flows poured from the rising Avalon block into that new basin, depositing the sediments that eventually became the Seboomook group of formations. Those deformed sediments exist along the western flank of the central slate belt and throughout the entire northern slate belt province.

Acadian Mountain-Building Event

The Acadian mountain-building event of Devonian time, 400 to 360 million years ago, happened after the last of the Iapetus Ocean floor sank though the oceanic trench and began its final plunge into the earth's mantle.

Maine has two belts of igneous rocks apparently related to the Acadian mountain-building event: the Piscataquis volcanic belt, exposed in the central volcanic province and in southeastern parts of the northern slate belt, and the coastal volcanic belt. Molten magma rose above the sinking ocean floor. Some erupted from volcanoes, more crystallized at depth to make masses of plutonic rocks, or plutons.

The two Acadian volcanic belts in Maine suggest two trenches, but it is not clear exactly how that might have worked. Some believe that trenches lay on opposite sides of the Iapetus Ocean and that its floor sank, eastward beneath the coastal volcanic belt and westward beneath the Piscataquis volcanic belt. Others suggest that the floor of the Iapetus Ocean sank beneath the coastal volcanic belt, and then

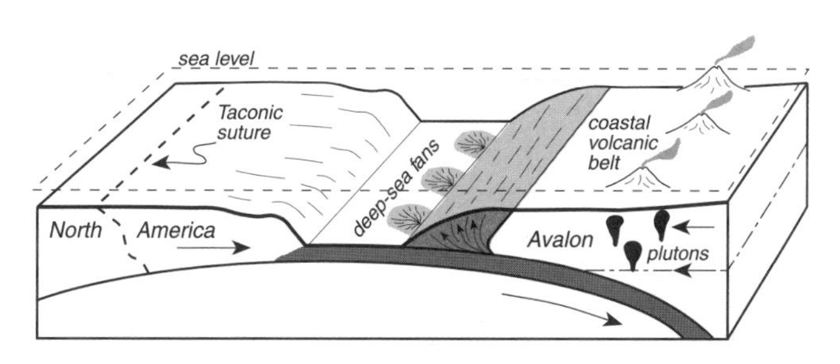

LATE SILURIAN TIME

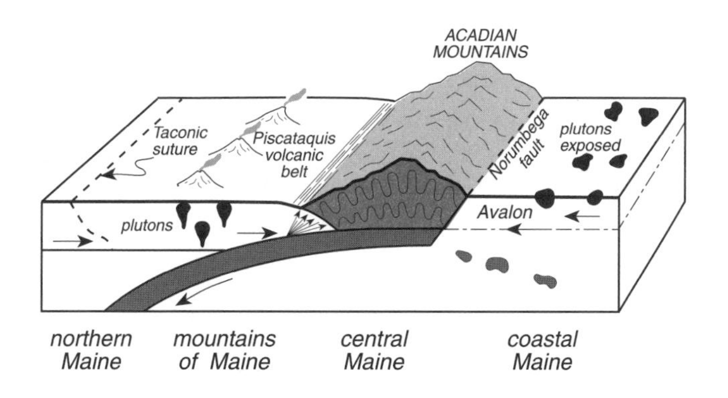

EARLY DEVONIAN TIME

Acadian mountain-building event. —Modified from D. C. Bradley, 1983

shortly thereafter the floor of the Connecticut Valley–Gaspé trough sank beneath the Piscataquis volcanic belt.

However it was accomplished, the Iapetus Ocean and all the other basins closed at the end of the Acadian continental collision. The sediments within them were crammed into the depths of oceanic trenches or compressed between colliding continental blocks. They heated and recrystallized, becoming new and different metamorphic rocks made of new and different minerals.

Most of the rocks of northern Maine never got very hot. They only partially recrystallized to become slates, which barely qualify as metamorphic rocks. The temperatures of metamorphic recrystallization generally increase southward and westward. Many of the rocks in southwestern Maine, and farther south, turned into streaky gneisses and flaky mica schists. They were so profoundly transformed that they hardly resemble the sedimentary rocks from which they formed. They are metamorphosed almost beyond recognition, especially close to the Sebago pluton.

The width of the belts of rocks folded and recrystallized during the Acadian event decreases toward the south, as if pinched between two fingers. That probably tells us something about the map outline of the colliding coasts of North America and Europe. Most of Maine formed in front of a large embayment of the Laurentian margin, just north of the large New York promontory. The promontory to the south took up most of the force of the collision, so the rocks there are tightly crumpled and much more metamorphosed. Meanwhile, the rocks in the bay were more gently compressed.

Although the increase in metamorphic grade farther south can be explained by greater compression along the New York promontory, some of this high-grade metamorphism is the result of a third mountain-building event, the Alleghanian.

Alleghanian Mountain-Building Event

The Alleghanian mountain-building event of 300 to 250 million years ago happened as the ancestral African continent collided with North America. The rocks directly involved are all south of Maine. Most rocks in Maine were already strongly deformed in the Acadian event, so the effects of the Alleghanian deformation are not easily observed. However, some of the faults that bound the Avalon portion of the coastal Maine province saw movement during the Alleghanian event. Some geologists argue that after it collided, the Avalon block slipped along the coast to its present position. (Geologist H. P. Woodward introduced the term *Alleghany* and intentionally spelled it differently than the Allegheny Plateau.)

Pangea and the Calamitous End of Paleozoic Time

After Europe collided with North America during the Acadian mountain-building event, and Africa collided with North America during the Alleghanian mountain-building event, most of the earth's inventory of continental crust was assembled into a single enormous continent. Geologists call it Pangea. It finally came together during Permian time, perhaps about 260 million years ago.

Then, at the end of Permian time, about 245 million years ago, the earth met a terrible calamity in which something more than 90 percent of all the kinds of animals then living vanished. The exact nature of the calamity is a matter of some considerable dispute. Quite a few geologists now suspect a connection between mass extinctions and eruptions of flood basalt lava flows—enormous eruptions hundreds of times greater than any that have happened since the human species emerged. Those geologists point out that the great Permian extinction happened while the Tungusska flood basalts of western Siberia were erupting. A few geologists argue that flood basalts erupt at the sites of great asteroid impacts. If they are right, then a massive impact caused the great Permian extinction, both directly and indirectly.

Whatever caused it, the great Permian extinction ended Paleozoic time. The zoo of animals that somehow survived it hardly resembled the much larger and more varied zoo that had lived before. But the plants fared better than the animals. Most of them survived into the next great era, Mesozoic time.

Mesozoic Time

As Mesozoic time started, the earth had plenty of plants but few animals, either on land or in the sea. The earth had the base of a food chain, but no food chain. The surviving animals quickly multiplied, diversified, and covered the earth with their progeny.

Geologists divide Mesozoic time into three periods: Triassic, Jurassic, and Cretaceous. Sometime during Triassic time, perhaps about 230 million years ago, the first dinosaurs appeared. Those early dinosaurs were free to roam across the great landmass of Pangea, with no oceans to block their way, and perhaps they did. Paleontologists who study their bones find that they hardly differ from Montana to Mongolia.

The Atlantic Ocean Opens

As Triassic time ended, Pangea began to break into great pieces that moved away from each other as new oceans opened between them. Those pieces are the modern continents, and the oceans between them are the modern oceans. The breaking that mainly concerns us here was the opening of the middle part of the Atlantic Ocean, which began about 200 million years ago.

When that great rift opened, it separated North America from Europe almost, but not quite, along the line of the vanished Iapetus Ocean. A few pieces that had been part of North America were left sticking to Europe, and a few that had been part of Europe were left clinging to North America. One of those was the Avalon terrane of eastern Maine.

Igneous rocks are the most visible evidence of the Mesozoic rifting in Maine. Stretching and cracking of the crust allowed basalt to rise from the mantle to the surface. Basalt dikes from this episode are common along the coast, but rare inland. The youngest rocks in Maine are small intrusions of gabbro, syenite, and diorite that lie along a curving line between York and Montreal.

It is hard to say much about the geologic history of Maine during most of the long period in which the Atlantic Ocean opened to its present width. No sedimentary rocks worth mentioning accumulated in Maine, or anywhere nearby, during those 200 million years. Those were mainly years of erosion, a condition in which the earth consumes its archives, instead of storing them.

The geologic record in Maine resumes with the great ice ages, which started sometime around 2 million years ago.

Ice Ages and Glaciers

Loose blocks of sandstone and slate litter the heights of Maine's highest peak, Katahdin. But the hard bedrock of Katahdin is granite, an igneous rock that in no way resembles sandstone or slate. Where did these foreign rocks come from, and how did they reach the summit, nearly a mile above sea level?

When C. T. Jackson did the first systematic geologic survey of Maine in 1835, such boulders were invariably classified as diluvium, the debris of Noah's flood. But in the 1840s the great Swiss naturalist Louis Agassiz began to argue that similar misplaced boulders in northern Europe had been dropped from great glaciers that had covered the entire landscape. He was by all accounts a charismatic speaker who packed lecture halls across Western Europe with people who paid good money to hear him tell of the great ice ages that had come and gone in the fairly recent past. He roused a furor wherever he went.

In 1847, Harvard University invited Agassiz to serve for a year as a visiting professor. He promptly learned English and continued to argue his theory of ice ages to large audiences, both on and off the campus. At the end of the year, the political situation in Europe combined with his own disorderly domestic arrangements to inspire Agassiz to remain permanently at Harvard. He distinguished that great university in many ways, for many years.

Agassiz never saw the great continental glaciers of Greenland and Antarctica, and could not have known much about them. Instead, he studied the much smaller glaciers of Switzerland and the deposits that the enormous glaciers of the ice ages had left strewn across Europe and North America. From them, he reconstructed in his mind the great glaciers of the ice ages and described them in words that still seem almost modern. Only a towering genius could look at messy glacial debris and at scratches on bedrock outcrops and see in them the signature of a sheet of groaning ice that covered a large part of a continent to a depth of thousands of feet.

While evidence of four to perhaps twenty glacial episodes is preserved elsewhere, the landscape of Maine contains a clear record of only the most recent ice age. It began about 40,000 years ago, reached a maximum about 18,000 years ago, and finally ended in Maine about 12,000 years ago. The glaciers of that ice age rather effectively bulldozed the record of their predecessors off the landscape, but their own record covers most of Maine, most obviously in all those boulders.

Glacial boulders in Smithfield. —F. J. Katz photo, U.S. Geological Survey

Boulders, Boulders Everywhere

Boulders that differ from the bedrock beneath are called erratics. The sandstone boulders that litter the granite summit of Katahdin are erratics; ice-age glaciers carried them there from outcrops somewhere northwest of the mountain.

Most of the boulders are in deposits of glacial till, sediment dumped directly from the ice. Glacial till is easy to recognize. If you see it in a roadcut, till appears as a disorderly mess of mud, sand, pebbles, and boulders, all mixed together. It looks like something a bulldozer must have scraped up and dumped.

Once the early settlers in New England cleared the land of trees, they found their fields covered with boulders. They dragged them to the edges of the fields to make stone walls that express the abundance of glacial till. Then the boulders returned, or more boulders appeared. Many people believed that they grew from seed rocks; if you could find and remove all the small seed rocks, they would not be able to grow to a size that could break a plow. More likely, frost heaving raises the new boulders to the surface.

Till along Sandy River in New Sharon.

Rocks conduct heat better than soil. When the ground freezes, ice forms first under the rocks and shoves them upward. Then, when the ground thaws, bits of soil settle under the rocks, preventing them from settling back whence they came. So every cycle of freezing and thawing raises the rocks in the soil a bit higher. They really do rise up out of the ground, just as in the worst nightmares of generations of farmers. Glacial till contains a seemingly endless supply of boulders that continue to rise as the seasons pass.

Moraines

Anything made of glacial till is a moraine. Glaciers simply plastered most of the till in Maine onto the ground, in what geologists call ground moraine. It tends to make undulating and rather lumpy surfaces littered with boulders. Most till is nearly impermeable to water, so ponds and marshes dot most expanses of ground moraine.

Where the margins of glaciers stood in one place for many years, the melting ice dumped deposits of till in long ridges that exactly record the position of the ice margin. Most morainal ridges are lumpy, with little ponds and marshes in the low areas. Boulders litter their surfaces.

The glacier of the last ice age advanced as far as southern New England and the margin of the continental shelf, where it left its terminal moraine about 18,000 years ago. That moraine now forms part of the south shore of Long Island, all of Block Island, Rhode Island, and the northern shores of Martha's Vineyard and Nantucket in Massachusetts. The glacier left few moraines after it retreated from southern New England, until it reached Maine.

Small end moraines were left along the Maine coast when the ice margin retreated through there. End moraines are not common in the rest of Maine, but they do exist.

Glacial Outwash

Ice-age summers were just as long as the ones we know, had just as many hours of daylight as ours, and were probably fairly warm. Those great glaciers shed enormous amounts of meltwater during the warm months of the year. And they must have shed meltwater in torrents while the glaciers were rapidly melting as the last ice age ended.

The melting ice liberated sediment, as well as meltwater. Most of that sediment was ordinary sand and gravel; some of it was rock flour, or finely pulverized rock. The sand and gravel was deposited in the beds of streams, in alluvial fans that built out from the front of the glacier, and in deltas in glacial lakes and in the seas that briefly cov-

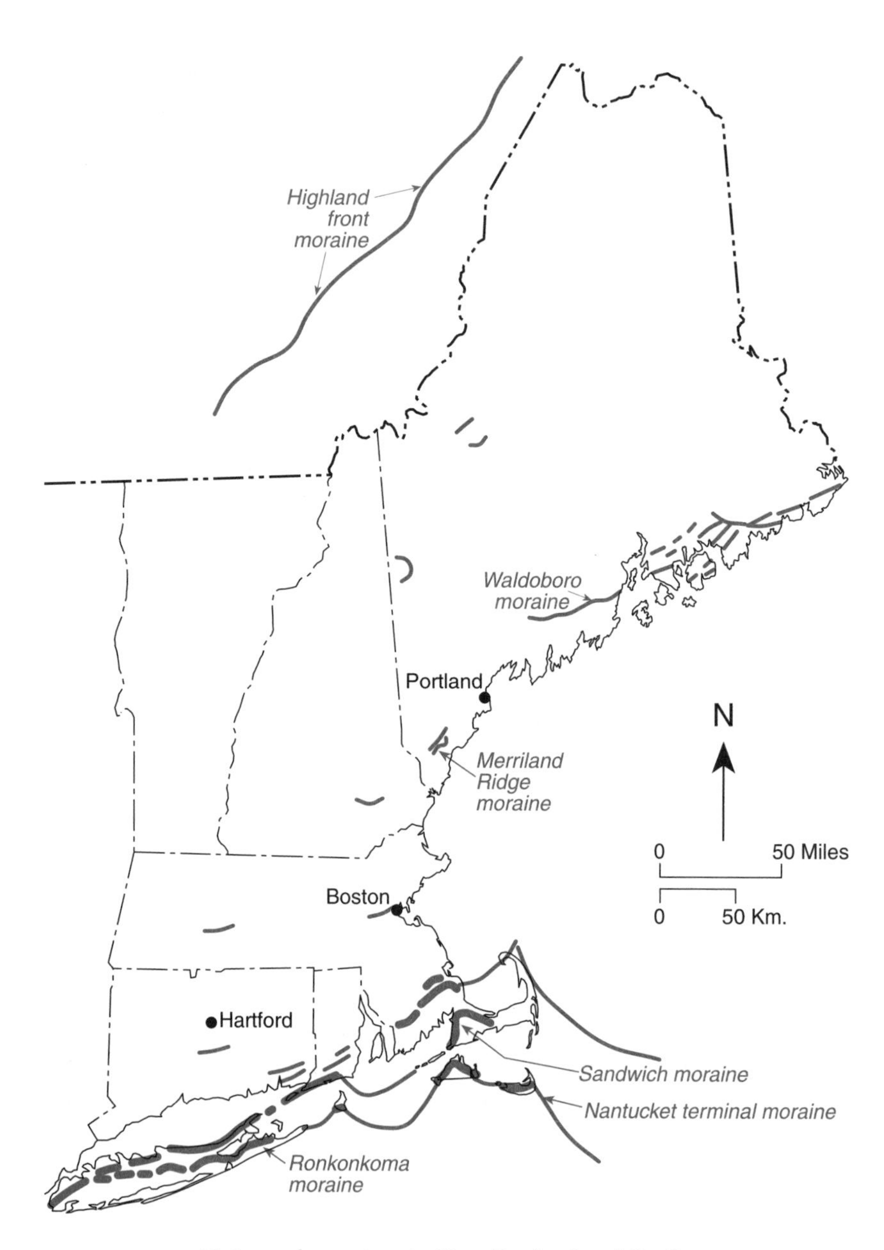

Major end moraines in New England and Québec.

ered Maine between about 14,000 and 12,500 years ago. The sand and gravel is called outwash. The rock flour usually settled only from water that had ceased flowing, in the deep parts of glacial lakes and on the seafloor.

Glacial outwash differs from glacial till in consisting of layers of sediment nicely sorted according to grain size: neat layers of sand and gravel. It looks very much like any other sediment deposited from a stream or in a lake. And the distinction between glacial till and glacial outwash is at least as important commercially as it is interesting scientifically.

Sand and gravel for construction aggregate are vital commodities in all populated areas. In many states, including Maine, sand and gravel mining is the most important mineral industry. Glacial till makes poor aggregate because it contains everything from clay to boulders, all mixed together. Glacial outwash, however, is an excellent source of aggregate because it contains sand and gravel neatly sorted into layers. A deposit of glacial outwash, however, near a market for construction aggregate is an extremely valuable resource, soon lost to the front-end loader and dump truck.

Eskers and Outwash Plains

Torrents of glacial meltwater roaring through winding tunnels and crevasses in the ice carry enormous volumes of sediment. Like any other stream, meltwater streams deposit some of their sediment load

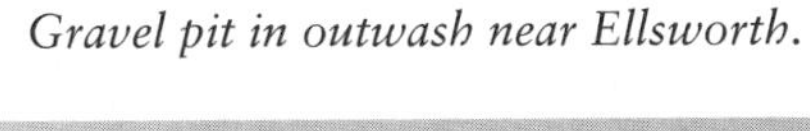

Gravel pit in outwash near Ellsworth.

in their channels. They flow through or beneath the ice on a bed of sand and gravel. When the last ice age ended, the melting glaciers left those stream sediments in low ridges that wind across the landscape. Geologists call them eskers, the technical term. Others are more likely to know them as horsebacks, an equally good name even though it does not appear in textbooks.

Eskers make wonderful sources of sand and gravel, especially because most of them are above the water table and therefore dry. The torrential currents that left these deposits washed away fine silt and clay particles. This left material that the construction industry calls clean gravel, something that does not have to be washed before it can be used. Most of the eskers in southern New England have been mined and hauled away. You can see them on old topographic maps, but not in the field.

The word *esker* comes from Ireland and describes the paths that often run along their tops. Many of the eskers in Maine cross long expanses of soggy ground moraine full of ponds and marshes. They provide a dry route through a boggy landscape, an obvious location

Esker ridge near Abol Pond and Abol Bridge, Baxter State Park.

for roads and houses. Those eskers will probably survive the demand for sand and gravel.

Maine has the longest eskers in the world. Some continue for more than a hundred miles, then end in broad outwash plains. You see in them the routes of meltwater streams that carried sand and gravel through the ice and through the moraine along its edge, then spread them out in glacial marine deltas. Very few eskers in Maine terminate above the former sea level.

Kettle Ponds and Kettle Lakes

The glaciers of the last ice age were still close to their farthest reach as recently as 15,000 years ago. They locked up enough water to drop sea level more than 300 feet below its present stand. Then the climate changed, and the glaciers melted so rapidly that by about 12,000 years ago they had disappeared from Maine. Sea level returned to nearly its present stand within a few thousand years.

Kettle ponds and kettle lakes are steep-sided, bowl-shaped depressions in glacial drift deposits. Blocks of ice that were buried in the drift as the glaciers retreated eventually melted, causing the land to settle in the shape of a kettle. Kettles often contain lakes or ponds because they extend below the water table.

Glacially Eroded Bedrock

At the maximum of the last ice age, a continuous glacier covered even the highest mountains in Maine. Later, after the ice sheet thinned, some of the higher mountains supported small glaciers of their own that poured down their valleys and into the regional ice sheet. That slowly moving ice left its mark in the eroded bedrock of Maine.

Glaciers quarry bedrock in some situations and abrade it in others. The two processes leave very different surfaces. Glaciers quarry bedrock mainly by freezing fast to its surface and then plucking blocks loose and carrying them away. Plucking happens mainly on surfaces that face in the direction of ice flow, on surfaces that face generally south. The melted ice leaves the quarried surfaces generally steep and raggedly irregular.

Glaciers use the sediment embedded in them to abrade bedrock surfaces; in effect, they sandpaper them. That happens mainly on bedrock surfaces that face into the direction of ice flow—on surfaces that face north. Now that the ice is gone, those surfaces are generally smooth, and they tend to have gentle slopes and smoothly sculptured forms. If the ice was carrying very fine sediment, it left a polished bedrock surface. But all glacially abraded surfaces have patterns of parallel scratches on them that precisely record the direction of ice flow.

Watch the bedrock of Maine for its evidence of glacial sculpturing on all scales, from little knobs to large hills. You will see a strong general tendency for gentle bedrock slopes to face north and for steep slopes to face south. And you will find that bedrock surfaces on the gentle north slopes are commonly smooth, and that patterns of parallel scratches or grooves striate them. The steep slopes that face south are ragged and lack glacial polish.

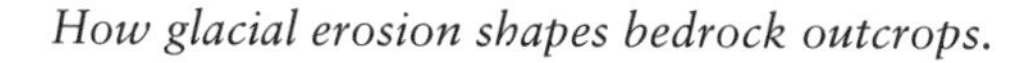

How glacial erosion shapes bedrock outcrops.

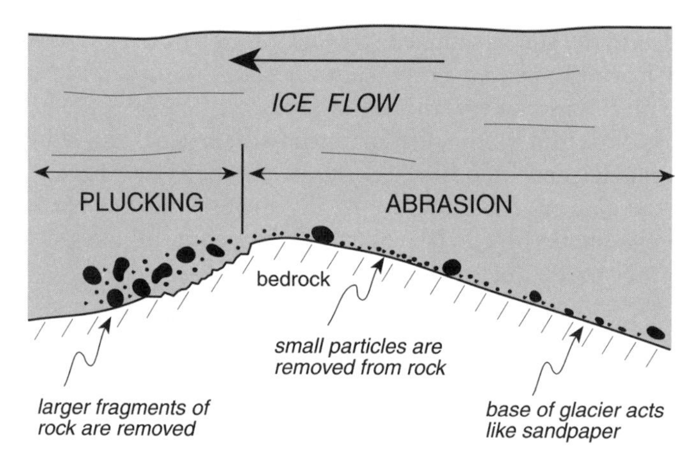

Abraded and plucked bedrock outcrop near Chesuncook Lake, Township 3, Range 12.

High-Water Deposits

About 17,000 or 18,000 years ago, all of Maine was beneath an ice sheet that flowed from central Canada to the New England continental shelf. When the glacier stood with its margin on the continental shelf, sea level was about 300 feet lower than it is now. The edge of the ice was on dry land—exposed continental shelf.

Meanwhile, the weight of the ice, which was more than a mile thick, had pushed the continental crust of coastal Maine below the ice-age sea level. As the ice age ended, sea level rose much faster than the earth's crust could float up. By the time the ice margin had retreated to the present Maine coast, and probably even before that, the rising sea caught up with it, floating the edge of the glacier. So coastal Maine was underwater and beneath a shelf of floating ice. And seawater flooded far inland up the major valleys.

In places where the edge of the ice was grounded, it deposited short lengths of glacial moraine, a type called DeGeer moraines. They typically consist of glacial till interlayered with outwash. Many were roughly handled as the ice, which was floating nearby, pushed across them.

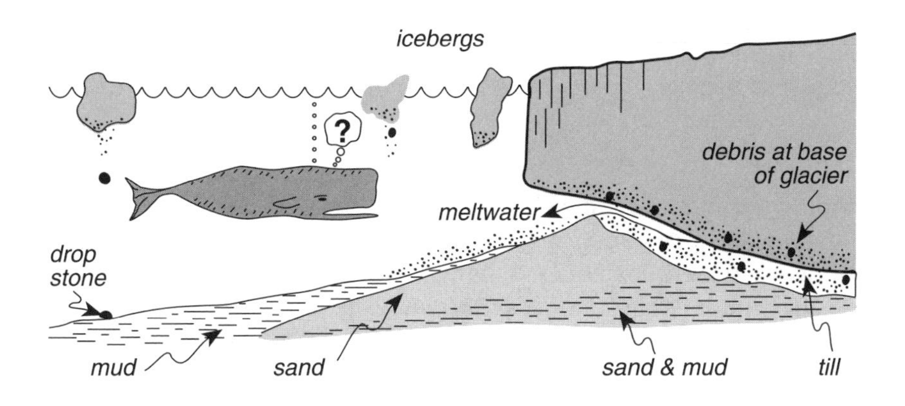

Formation of DeGeer moraines. —Modified from M. J. Retelle and K. M. Bither, 1989

The meltwater streams that flowed in tunnels under the ice emptied onto the coast. There they dropped their load of glacial outwash sediment to deposit deltas at least partly submerged beneath the shelf of floating ice. Most of the long eskers end in these deltas.

Glacial meltwater dropped mud in the shallow coastal water, where it settled to the bottom to make large deposits, the Presumpscot formation. The salt water coagulated the mud, making it settle rapidly

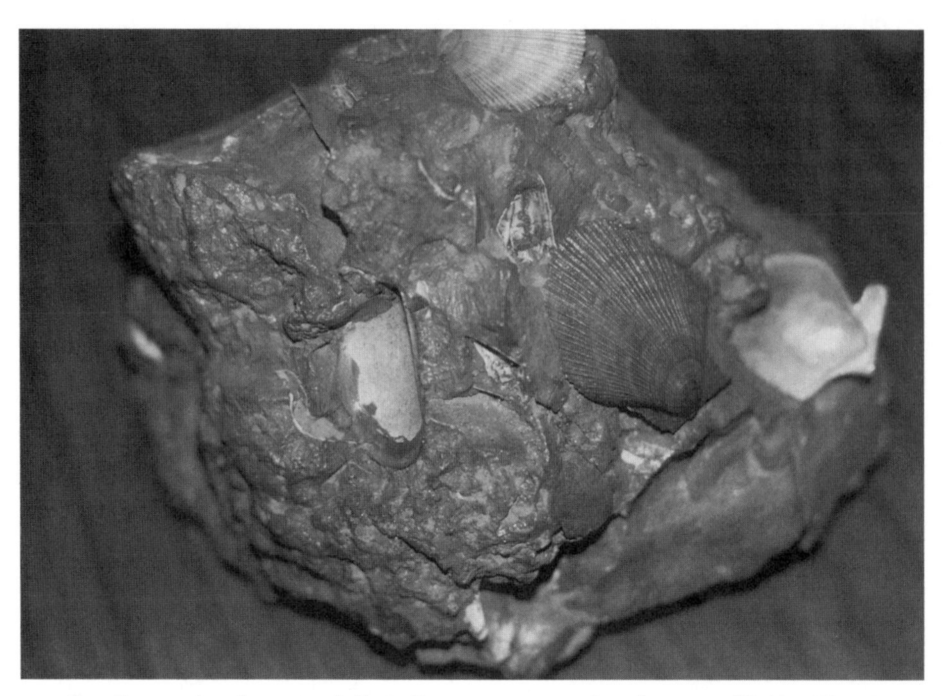

Fossil mussel and pecten shells in Presumpscot marine clays near Union. These shells are about 13,000 years old and are in the Maine Geological Survey museum in Augusta.

without forming prominent layers. The Presumpscot formation is noted for its wide variety and number of fossils, remains of animals and plants that lived near the ice margin as the mud accumulated. The most common types of fossils are clams and snails. Rarer fossils include driftwood, the skeletons of whales, and the tusks of walrus.

Numerous radiocarbon datings make the Presumpscot formation one of the most completely dated in the world. This work shows that the ice margin was near the present coastline between 13,800 and 13,500 years ago.

By 13,000 years ago, the ice had retreated to near the limit of seawater submergence, in places as much as 100 miles from the modern coast. Meanwhile, the continental crust was floating up as the ice melted. The rise of the land made the sea retreat. By about 12,500 years ago, Maine was once again above sea level. By about 11,000 years ago the land was about 150 feet higher above the sea than it now is. Then as ice continued to melt, sea level rose until it reached close to its present stand 2,000 years ago.

In the nineteenth century, nearly every town had a pit where clay from the Presumpscot formation was burned to make bricks for chimneys and for the homes of the wealthy. Those kilns are long gone, but modern brick factories still operate near Lewiston and Portland.

The flooded parts of Maine about 13,000 years ago. The shaded areas were above the water.

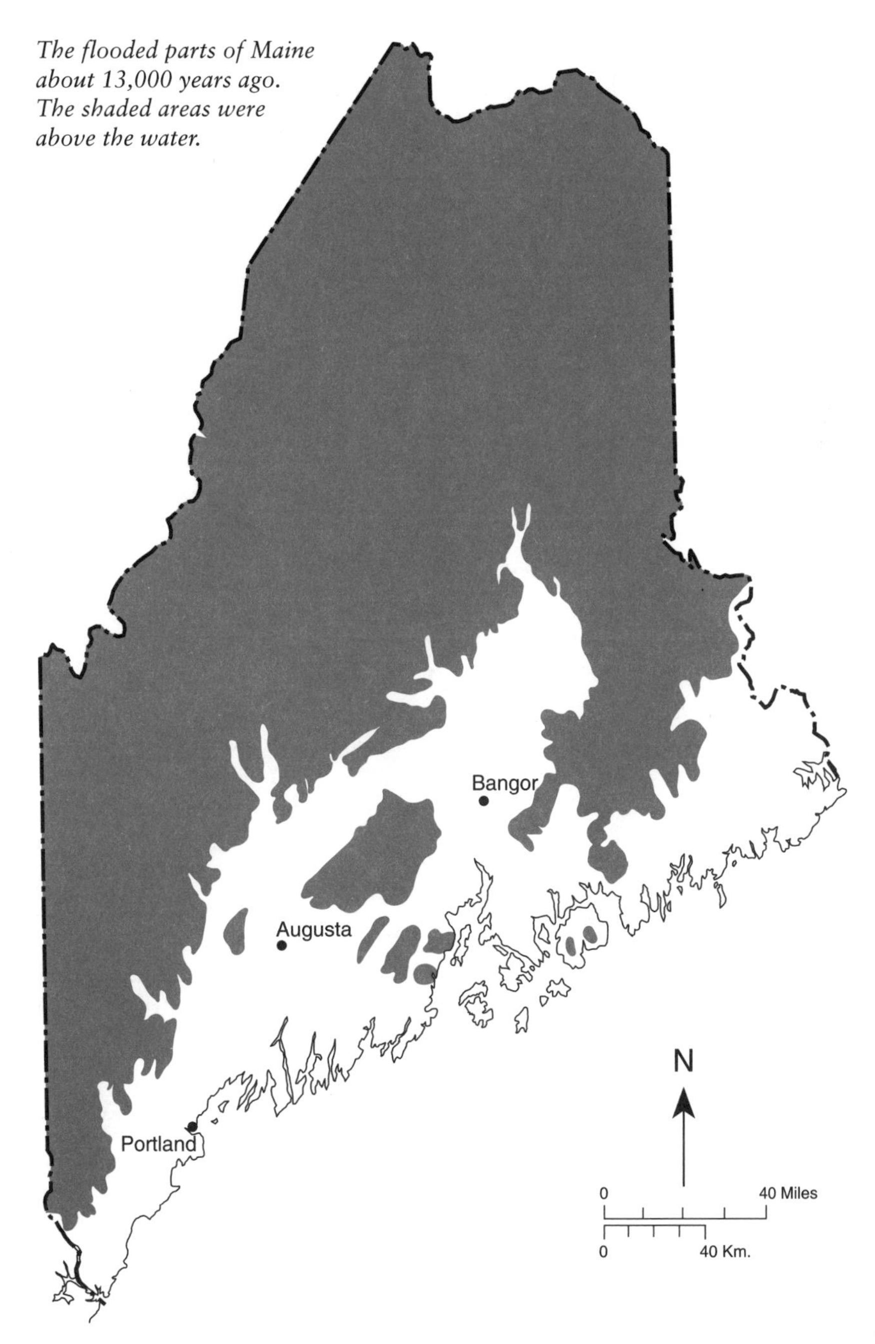

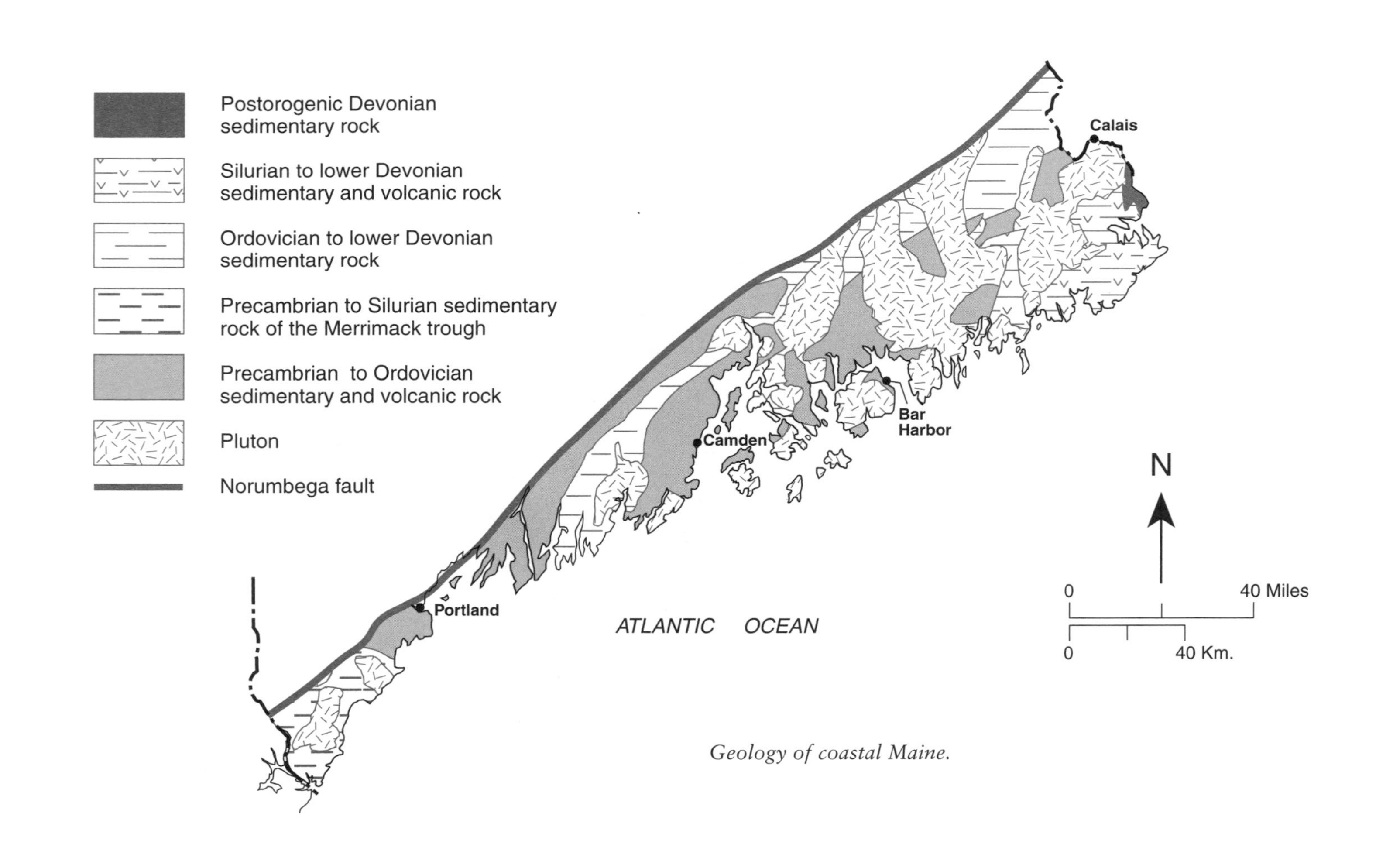

Geology of coastal Maine.

Coastal Maine

Just as most visitors to Maine still do, early explorers from Europe stayed near the coast. Norsemen evidently sailed down the coast of Maine between 1003 and 1011 but left no record. In 1497 the explorer John Cabot and his sons made a map of the East Coast from Newfoundland to the Carolinas, upon which the English based their claim to North America.

Three English sailors who survived an ill-fated expedition to Mexico appear to have been the first white men to spend any time in Maine. They walked from the Gulf of Mexico up the East Coast and across Maine, arriving in about 1568. They reached a French colony at the mouth of the St. John River in what is now New Brunswick and finally sailed for home on a French trading boat. Except for this incredible journey, it was the French who most often visited Maine in the century after the Cabots.

Samuel de Champlain and his patron, Pierre du Gaust, the Sieur de Monts—literally Sire of the Mountains, or Sir Mountain—established the first European colony in Maine on St. Croix Island, near the mouth of the St. Croix River, in 1604. This party spent the winter on the island and were the first to celebrate Christmas in New England; they abandoned the settlement the following spring. Champlain's exploration of the coast of Maine and Atlantic Canada was the basis of the French claim to these lands, which they called *l'Acadie*. The dispute with the British over these lands continued for about 150 years and through five French and Indian Wars.

In 1607, the English founded their first colony in New England at the mouth of the Kennebec River at Popham. It was also soon abandoned, but not before the first sailing vessel made in North America, the *Virginia,* was launched. The arrival of the *Mayflower* in 1620 firmly established English settlement of coastal New England, although French missionaries and traders continued to maintain small outposts near Indian villages, mostly in interior Maine and east of the Penobscot.

The English control of Maine was furthered by the Council of New England, which established the Great Patent of New England; a patent, in this sense, means a grant of public lands, or the granted lands themselves. In 1622, the council granted Sir Ferdinando Gorges and Captain John Mason the land between the Merrimac and the Kennebec Rivers, to be known as the Province of Maine. In 1629, the men divided their land at the Piscataqua River, with Gorges getting the Maine portion. In 1630, much of Maine east of the Kennebec River was granted to men from London and Boston, the land becoming known as the Muscongus Patent. These lands became the Waldo Patent in about 1750. Samuel Waldo was a general in the militia, composed mostly of Mainers, that defeated the French in Nova Scotia in 1745.

Maine became part of one of the thirteen colonies and states, as the Province of Maine, a part of the Commonwealth of Massachusetts. It became a state in 1820 as part of the Missouri Compromise. The Webster-Ashburton Treaty established the present northern and eastern boundaries of Maine in 1842. York and Kittery are the two oldest towns in Maine, dating from 1642 and 1647.

Rocks of the Maine Coast

The kinds of rocks in the coastal part of Maine differ from those in other parts of the state. They are generally older and contain different fossils. This is a distinct geologic province, the Avalon terrane. It is a continental mass that collided with the eastern margin of North America during Devonian time, perhaps sometime around 400 million years ago. The Avalon terrane was the last big piece added to Maine. It was named for the Avalon Peninsula in Newfoundland.

The Norumbega fault separates the Avalon terrane from the rest of Maine. It is both a fault and a continental suture, the zone along which the Avalon terrane welded itself to North America. Geologists trace the Norumbega fault, under different names, from New Brunswick to Connecticut.

Some of the rocks in the Avalon terrane are older than any others at the surface in the state, except those of the Chain Lakes massif in northwestern Maine on the border with Québec. The Grenville basement rocks that probably lie deeply buried beneath northern Maine are also older than those of the Avalon terrane.

Rocks of the Avalon terrane have a history unlike that of rocks across the Norumbega fault. That leads geologists to suspect that they were nowhere near the rocks north and west of the Norumbega fault during its active history. The oldest rocks in coastal Maine are mostly volcanic rocks that erupted during very late Precambrian time, and sedimentary rocks composed of material eroded from them. They grade upward into slates and sandstones that were deposited during early Paleozoic time.

Long before geologists developed their ideas of plate tectonics, they recognized that the early Paleozoic rocks south and east of the Norumbega fault contain fossils more like those of northern Europe than those in western New England. It is now clear that the Avalon terrane was part of Europe during early Paleozoic time, while the rest of New England was on the other side of the Iapetus Ocean. The two sets of fossils were not close to each other until the Iapetus Ocean closed and North America welded to Europe during Devonian time.

Acadian Mountain-Building Event

A discontinuous belt of volcanic rocks that erupted during Silurian and early Devonian time extends from coastal Massachusetts, through the Mount Desert Island region, to easternmost Maine near Eastport. This is the coastal volcanic belt. More igneous rocks, both plutonic and volcanic, of Devonian age underlie a much larger area of central Maine. Together these igneous rocks are part of the evidence for the Acadian mountain-building event.

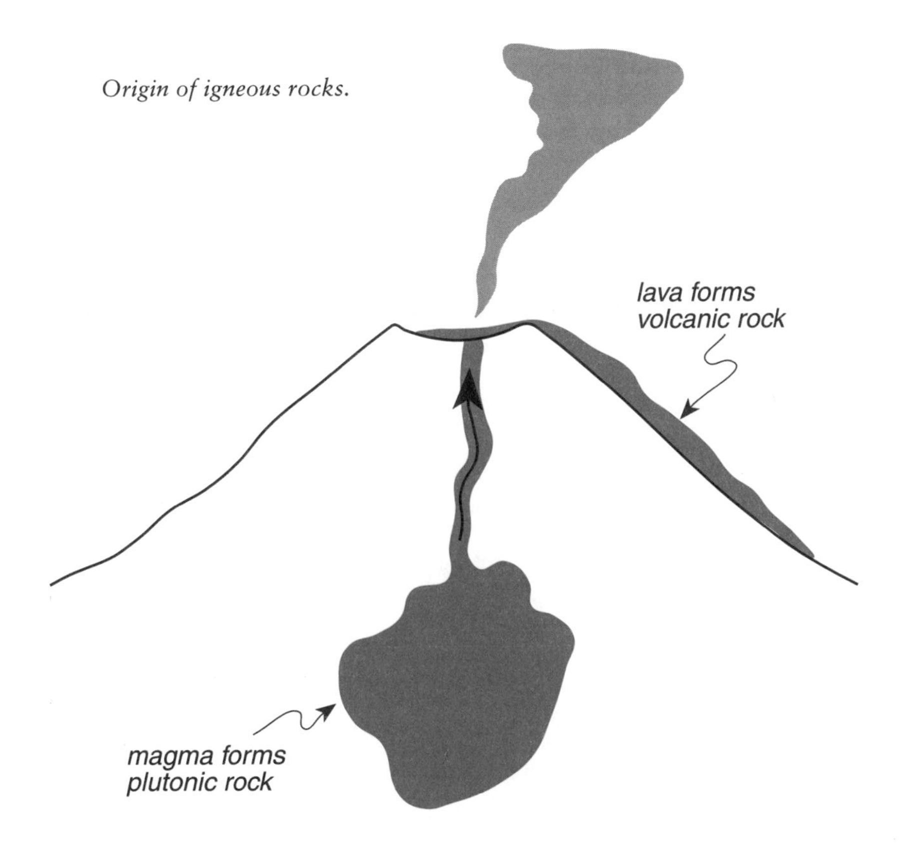

Origin of igneous rocks.

A number of tectonic models have been proposed for the Acadian mountain-building event. The most satisfying of them proposes that ocean floor was then sinking southeastward under the Avalon terrane. An oceanic trench developed where the sinking oceanic plate bent down to start its long plunge into the mantle, and a chain of volcanoes grew along a line parallel to it. The abundant coastal plutons were the magma chambers from which the volcanoes erupted. Many of them eventually rose high enough to invade their own volcanic piles. That, and the very long time it may take for a pluton to cool, explains why they tend to be younger than the volcanic rocks that erupted from them. The mostly granite plutons from Penobscot Bay eastward to Eastport are known as the Bays of Maine igneous complex. And complex they are. The plutons in this area were still hot when they were intruded by basalt. Broken basalt dikes and pillowlike features are common in this area.

Across the Norumbega fault from the Avalon terrane is a vast thickness of mudstones that were deposited on the deep ocean floor during Silurian and Devonian time. These rocks were intensely folded as the ocean floor beneath them sank under the Avalon terrane, rudely cramming them into the oceanic trench.

Squeezing the sediments crumpled them into a tightly folded mass wedged between the Avalon terrane and the margin of the North

Broken basalt dike in Meddybemps pluton in Alexander.

American continent. Finally, the ocean floor stopped sinking through that trench, and the active volcanoes were snuffed out.

Meanwhile, or perhaps just afterward, the line of the oceanic trench shifted, and oceanic crust began to sink beneath north-central Maine. Another, but smaller, package of mudstone was crammed into this later version of the trench, and another chain of volcanoes erupted parallel to it. That was the final stage of the Acadian mountain-building event, at the end of early Devonian time, about 380 million years ago.

Nearly all the bedrock of Maine was then in place. North America was attached to Europe, as part of the supercontinent Laurasia. In late Carboniferous time, about 300 million years ago, Africa collided with North America along a line from southern New England to Alabama, raising the central and southern Appalachian Mountains. This was the Alleghanian mountain-building event. A few features in coastal Maine that appear to be younger than the Acadian mountain-building event may date from the Alleghanian event. For example, the Norumbega fault broke some late Devonian and early Carboniferous plutons in movements that may have occurred during the Alleghanian mountain-building event.

Metamorphism in Coastal Rocks

The sedimentary rocks between Kittery and Portland and between Ellsworth and Calais show remarkably little effect of metamorphism, considering what they have been through. Between Brunswick and the Penobscot River similar rocks are highly metamorphosed. Also, in much of this central region of coastal Maine, the rocks south of the Norumbega fault are more strongly metamorphosed than those to the north. It has been suggested that the highest levels of metamorphism in Maine are related to the heat from the Sebago pluton, which was emplaced in Carboniferous time. The rocks now between Brunswick and the Penobscot River may have been in southwestern Maine when they were heated by the Sebago intrusion. Sometime later, movement along the Norumbega fault system shoved these high-grade metamorphic rocks eastward down the coast. A small sliver of these high-grade rocks moved even farther east, beyond Bangor, where it slid next to another sliver of sedimentary rock that was not metamorphosed.

Opening the Atlantic Ocean

During late Paleozoic and early Mesozoic time, almost all the earth's inventory was assembled in a single great continent, Pangea. Maine

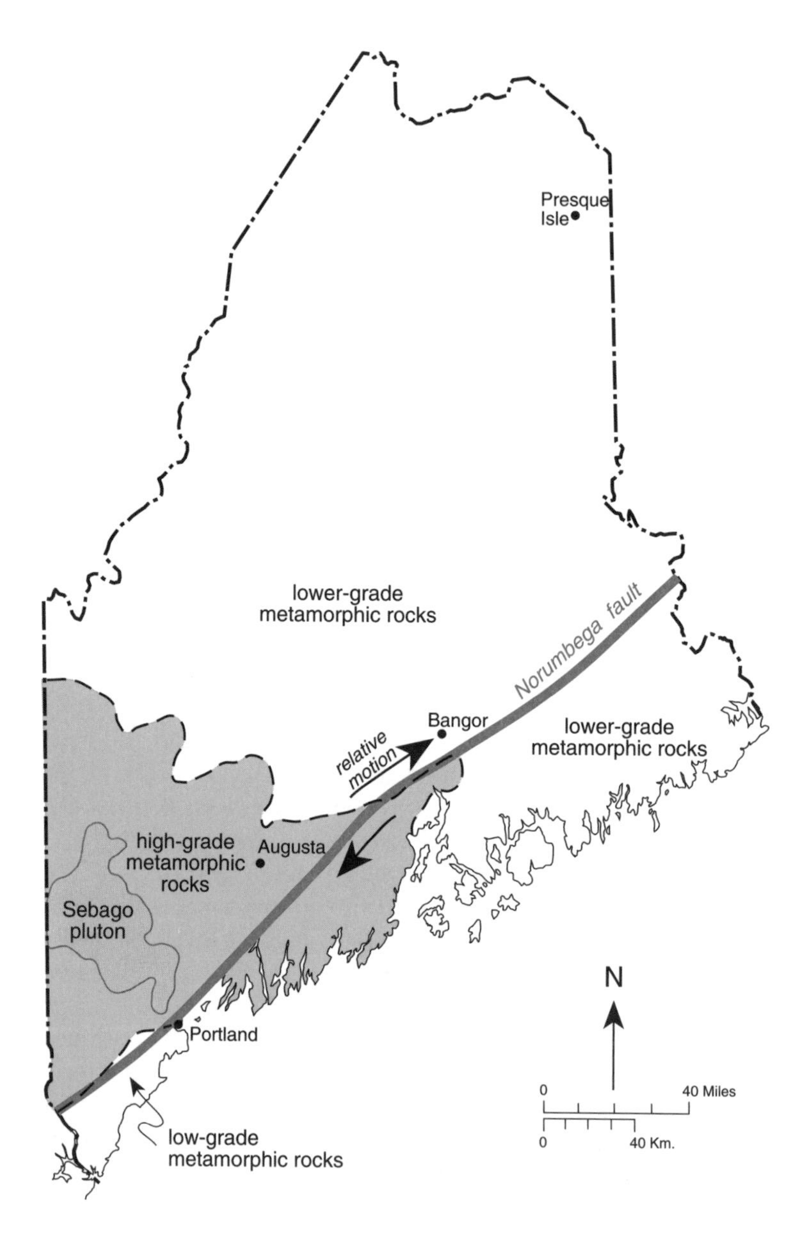

Displaced metamorphic zones near the Norumbega fault.

was deep in its interior. You can reconstruct its position by cutting the Atlantic Ocean out of a map and then moving North and South America against Europe and Africa. Maine would be thousands of miles from the coast, in any direction.

Sometime during Triassic time, more than 200 million years ago, the central part of the Atlantic Ocean began opening between the east coast of North America and the coasts of Europe and North Africa. In its first stages of opening, the new Atlantic Ocean probably looked about as the Red Sea does now. It has been growing wider ever since, at an average rate of approximately 2 inches per year, about as fast as your fingernails grow.

The first stages of that rifting were a bit tentative; the continents did not simply crack apart and start moving away from each other. Instead, a number of rifts formed and opened a short distance before one established its dominance and began to develop into the Atlantic Ocean. Those failed rifts filled with red muds and sands that contain the bones and footprints of dinosaurs. The Bay of Fundy and part of the Gulf of Maine formed in that way, as did the Connecticut Valley. If the Connecticut Valley rift had not failed, and had the Atlantic opened along it, Maine, New Hampshire, Rhode Island, and eastern Connecticut and Massachusetts would now be part of southern Europe. Other such failed rifts, filled with red mudstone and sandstone, exist on the eastern side of the Atlantic Ocean.

While those rift valleys were dropping and filling with sediment, molten basalt magma was rising from the earth's mantle through long fractures in the crust. Some of that magma erupted as great lava flows that poured into the dropping rift valleys, where we now see them interlayered with the red sediments. And some of that magma crystallized within the fractures to make dikes. They were, in all likelihood, the plumbing for big basalt flows now long since eroded off the landscape except in the failed rifts. The basalt dikes and flows in the northeast are lower Jurassic in age. Where preserved, sandstones and shales that overlie the basalt flows are dimpled by the tracks of dinosaurs.

The basalt dikes average about 3 feet across, and you can commonly trace them hundreds of feet if the exposure is good. If exposure were better than it is, you could probably trace them for miles. A huge dike, 60 to 90 feet wide, has been traced more than 100 miles, from near Yarmouth through Bristol to Pemaquid and beyond. Dikes with similar chemistry and orientation are found in Connecticut and New Brunswick and may be continuations of the giant dike in Maine. All of these dikes presumably continue down to the base of the continental crust, 30 miles or so. Basalt dikes are common near the coast

Mesozoic basalt dikes on I-95 in York. They intrude the possibly Triassic-aged Agamenticus igneous complex.

of Maine, but rare inland, away from the area of continental splitting. All trend northeast, generally parallel to the coast.

The Atlantic Ocean opened along a line close to the one along which the Iapetus Ocean closed 150 million years earlier. But the coincidence of oceans was not perfect; some pieces that had been part of Europe when the Iapetus Ocean existed wound up in North America after the Atlantic Ocean opened, and vice versa. New England and the Maritime Provinces contain areas in which the Paleozoic fossils are the remains of animals that then lived in Europe.

The White Mountains Igneous Rocks

The youngest rocks in Maine are a series of small igneous intrusions of varied composition. Most are younger than the coastal dikes; they range in age from late Triassic to late Cretaceous, from about 200 million years to about 70 million years. Geologists assign them to the White Mountains magma series.

Geologic maps show these intrusions arrayed along an arc that extends from Montreal to the coast and beyond. They become progressively younger to the east. Many geologists interpret them as the track of a hot spot, a small area of abnormally hot rock in the uppermost part of the earth's mantle that generates volcanic activity directly above it. As the lithosphere plate moves across the hot spot, which is stationary in the mantle, volcanoes appear one after the other along a line—on a map, a dotted line. This chain of igneous intrusions ends in a chain of extinct undersea volcanoes called the New England seamounts, which continues almost to the middle of the Atlantic Ocean. An active undersea volcano presumably marks the present position of the New England hot spot.

The line of small Mesozoic plutons that mark the passage of eastern North America over a hot spot. In general, these rocks are younger the closer they are to the coast.

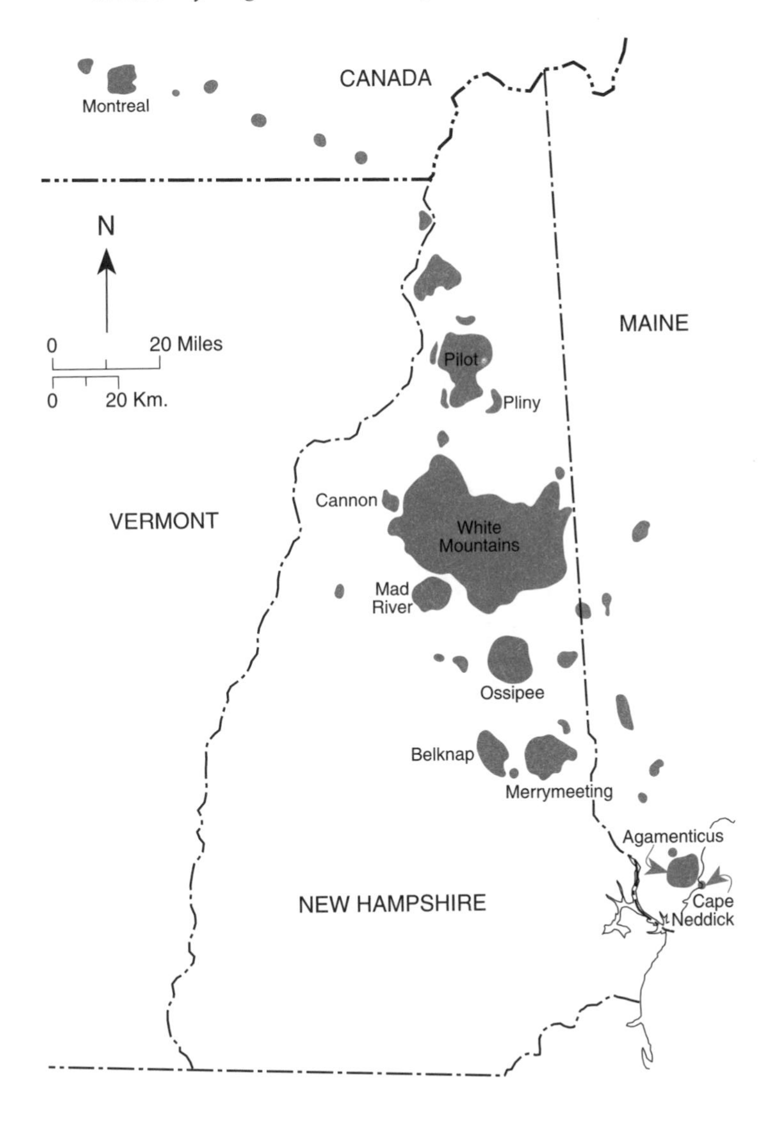

Coastal Glacial Deposits

As recently as about 18,000 years ago, during the last of many great ice ages, a vast ice sheet that extended from northern Canada to the New England continental shelf covered all of Maine. When the glacier stood with its margin on the continental shelf, sea level was much lower than now, maybe 200 or 300 feet lower, because the glaciers contained so much water. The ice sheet stood on dry land. The ice age came to a maximum sometime around 18,000 years ago. Then the climate changed, and the ice began to melt—rapidly. The sea level began to rise as meltwater ran into the ocean.

The great weight of the ice, more than a mile thick, had pushed the crust down below the low sea level of the ice age. When the ice melted, sea level rose much faster than the earth's crust could float back up. By the time the ice margin had melted back to the present Maine coast, and probably before, the rising sea caught up with the retreating ice front and flooded the depressed crust of coastal Maine. The edge of the glacier floated, a shelf of ice floating over coastal Maine. Great meltwater streams flushed sediment shed from the melting ice onto the shallow seafloor of coastal Maine.

The glacial deposits laid down in shallow seawater include moraines, deltas, and mud. The most common glacial marine moraines occur in low, short segments that were laid down where the ice was grounded. These small features, the DeGeer moraines, consist of layers of sand, gravel, mud, and till, commonly deformed by ice push. Larger glacial marine moraines are really deltas deposited next to the edge of the ice. Large deltas record the sea level at the time of their deposition; they differ from the smaller moraines mainly because the ice margin stayed in place longer, allowing the sediments to build above sea level. Glacial marine mud, the Presumpscot formation, consists mostly of finely ground particles of rock carried to the sea by meltwater. Upon reaching the salt water, the mud settled quickly to the seafloor without forming prominent layers.

Dance of Land Elevation and Sea Level

Age dating shows that the ice margin stood near the present coastline about 13,500 years ago. By 13,000 years ago, the ice had melted back to near the limit of marine submergence, as much as 100 miles from the coast. Meanwhile, the crust, unburdened of ice, was floating back toward its normal elevation.

As the land rose, the sea retreated. By about 11,000 years ago, coastal Maine was once again above sea level, about 200 feet higher above sea level than now, because the land was then rising faster than

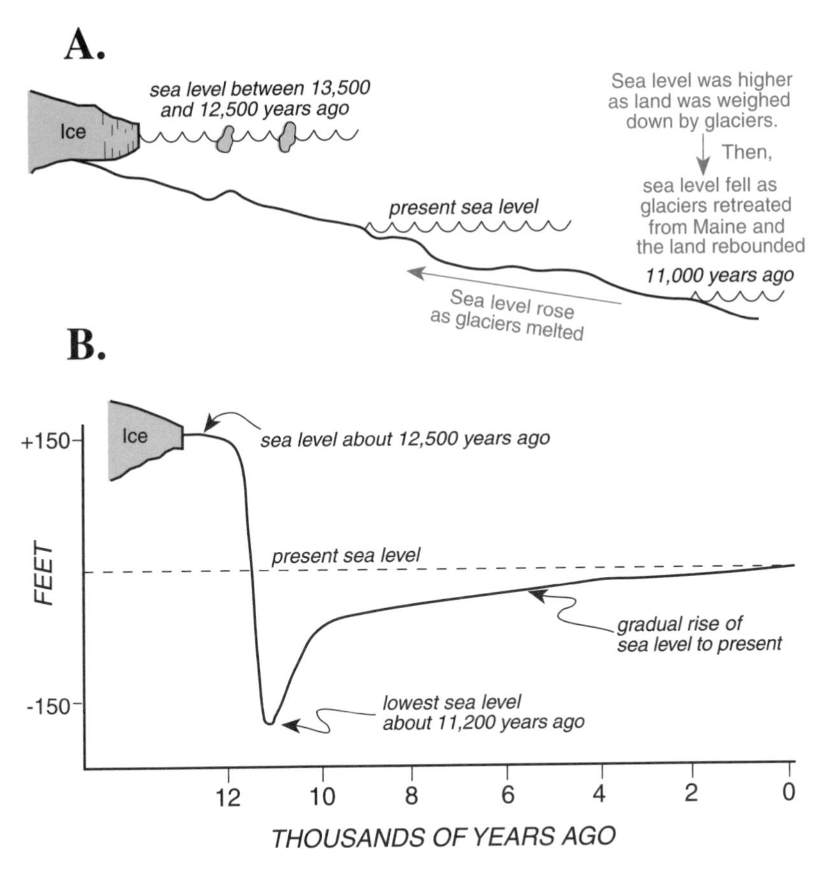

Sea level changes along the Maine coast during the last 13,000 years.

meltwater was raising sea level. Then sea level rose rather quickly along the Maine coast as the land elevation stabilized and melting glaciers continued to pour water into the ocean. By about 2,000 years ago, sea level was near its present stand along the coast of Maine.

In Maine, as elsewhere, deltas record the elevation of the water in which these sediments were left, even though the water drained from the land thousands of years ago. The sloping layers in these deposits were formed by sand and gravel sliding down the front of the delta. They are called the foreset beds. As more sediment was carried to the delta front, new foreset beds built into the water. The flatter beds were formed in outwash stream channels that carried sand and gravel to the delta front over the old foreset beds. They are the topset beds. The old water level where a delta formed is shown by the contact

Topset, foreset, and bottomset beds in a glacial marine delta in Washington.
—Woody Thompson photo, Maine Geological Survey

between the foreset beds and the topset beds. In some pits you can see bottomset beds, fine sand that flowed out onto the seafloor in front of the delta.

You might expect deltas deposited in seawater to have the same elevation at the intersection of their foreset and topset beds. Not so. They become higher toward the northwest at a rate of about 2 feet per mile. In the Connecticut River valley, deltas formed in a long glacial lake there become higher to the northwest by about 4 feet per mile. That lake maintained a stable elevation while all the deltas were formed. Along the Maine coast the story is different. One line of deltas was formed along the ice margin. By the time the ice had retreated a distance north and stopped to form some new deltas, sea level rose, about 2 feet for every mile that the ice front retreated. Those deltas formed at a higher elevation than the first set. This becomes obvious when you examine these deposits, because some were flooded by the higher seas and reworked by waves. Then the earth's crust tilted up to the northwest as it rose. That happened because the ice was thicker toward the northwest, so the crust floated higher after it melted.

Interstate 95 (Maine Turnpike) and U.S. 1
Kittery—Portland
43 miles

U.S. 1 follows part of one of the oldest roads in Maine, the King's Highway, which was established in the late 1600s to connect Boston and Machias. In the beginning, the King's Highway was little more than some markers placed along old Indian trails. Some sections were former mastways, where long pine for masts was carried to water. Towns made their own improvements.

By 1913 the Maine Highway Commission was established to oversee many of the roads. The federal Department of Transportation established U.S. 1 as a numbered highway in 1925. The prominent historian and critic Bernard DeVoto once condemned this section of U.S. 1 as "full of drive-ins, diners, souvenir stands, purulent amusement parks, and cheap-jack restaurants," passing it all off as a "neon slum." This upset many a Mainer, not the least being the late E. B. White, writer for the *New Yorker* and author of the children's classics *Charlotte's Web* and *Stuart Little*, who maintained a summer and then retirement home in Brooklin. White defended U.S. 1, saying, "You can certainly learn to spell 'moccasin' while driving into Maine, and there is often little else to do, but steer and avoid death." The Kittery to Portland segment of the Maine Turnpike, now I-95, was opened in the fall of 1947, somewhat earlier than planned, to allow quicker access to the widespread forest fires that plagued the state that fall, particularly in the Sanford area.

Thick glacial deposits laid down in shallow seawater cover nearly all the bedrock along these roads, giving a poor introduction to the bedrock geology of the state. Some of the best bedrock exposures in this part of Maine are in sea cliffs, with numerous outcrops of plutonic rocks ranging from Devonian to Cretaceous in age. Large glacial sand plains dominate the landscape. After sea level rose at the end of the last ice age, waves reworked these plains to make the largest beaches in Maine. You can reach them from the shore roads off U.S. 1.

Most of the layered rocks of southwestern Maine are classified into groups or sequences of formations. Many of the formations resemble one another so closely that they look alike to the untrained eye, and to many trained eyes. A former state geologist passed them off as "coast rocks." The general lack of fossils makes it difficult to assign ages to these rocks. Igneous intrusions help.

FORMATIONS:

cu Cushing
ce Cape Elizabeth
k Kittery

K Cretaceous pluton
t Triassic pluton
c Carboniferous pluton
d Devonian pluton
Fault
Geologic contact

NORUMBEGA FAULT ZONE
spread footing at exit 6A overpass in Scarborough
remains of ice-age mammoth found in Scarborough
Portland
Old Orchard Beach
Swenson Pink granite quarry
Merriland Ridge moraine
ATLANTIC OCEAN
Wells Beach
Ogunquit Beach
Bald Head Cliff
black granite quarry in South Berwick
Cape Neddick layered gabbro
eroded drumlins in Eliot
Kittery
Swenson Green granite quarry
NEW HAMPSHIRE
N
0 5 10 Miles
0 10 Km.

Geologic features of the region between Kittery and Portland.

It is generally possible to obtain an age date on a pluton that intrudes one of these enigmatic rock units. That, of course, directly reveals the age of only the igneous rock, which must be younger than the enclosing sedimentary rocks. In some cases faults break dated plutons, and in other cases the pluton cuts unbroken across the fault. Those relations establish a youngest or oldest possible age for the fault. In every case, a fault must be younger than any rock it breaks. Faults may help establish an age for the sedimentary formations through their involvement with dated plutons. Age relations so laboriously inferred suggest that the major structures in the sedimentary formations of the coastal region date from the Acadian mountain-building event. Some of the major faults may have moved since then.

Eroded Drumlins in Eliot

North from the high bridge that connects New Hampshire and Maine you can see several low, rounded hills; the largest is Great Hill in Eliot. These are glacial drumlins, maybe a dozen of them. A number of years ago while working in the area I discovered that these were not ordinary drumlins. Instead of being solid glacial till the way they are supposed to be, these have sand and gravel deposits draped over their northern ends. This sand and gravel was well known to the gravel operators, who seem to be able to smell it a mile away and are usually well on their way to removing it from the face of the earth before geologists get wind of it.

These drumlins were washed by waves in the shallow sea that flooded coastal Maine during the retreat of the last glacier, about 13,000 years ago. The till in the drumlins contains just about any size rock fragment that you can imagine. The waves removed the sand, gravel, and mud and left the big rocks behind. The sand and gravel collected in the calm waters on the north sides of the drumlins, and the mud settled out in deeper water as part of the Presumpscot formation.

Merrimack Group

The Merrimack group of formations extends from Biddeford southwest into New Hampshire and Massachusetts. Geologists believe it dates somewhere between Ordovician and Silurian time. Most of the Merrimack rocks are mudstones with sedimentary features that indicate derivation from an easterly source, using present compass directions. A common member of the Merrimack group along this stretch of coast is the Kittery formation, a prominently layered, multicolored mudstone. The lighter beds are sandy, with the overlying mud layers

Mudstone in the Kittery formation on Bald Head Cliff in York. The diagonal line of white blobs is quartz veins intruded along faults.

varying from gray to tan to purple. The lighter colored mudstone may be derived from volcanic ash, erupted somewhere nearby.

Casco Bay Group

The Casco Bay group of rock formations extends from Saco, through Portland, and down the coast to Muscongus Bay near Friendship. It is more strongly metamorphosed to the northeast than is the Merrimack group. Geologists believe that these rocks date to Ordovician time. The rocks are now schists, marbles, quartzites, and volcanic rocks. Some were originally mudstones, and even though they are now metamorphosed almost beyond recognition, it is still possible to infer their origins. Faults bound many of the rock units within the group.

Mesozoic Dikes and Hot Spot Volcanic Rocks

A group of whales is a pod, a group of quails is a covey, a group of tourists is a passel, and a large number of basaltic dikes is a swarm.

Between Kittery and Ogunquit is a swarm of Triassic and Jurassic dikes. They formed as basalt magma filled fractures that opened during the rifting that started the Atlantic Ocean. These dikes cut through all sedimentary and plutonic rocks in the area except the Cape Neddick gabbro, which is a younger, Cretaceous pluton associated with the New England hot spot.

The Agamenticus igneous complex is a large, nearly circular area in which several kinds of granite and syenite intruded one another. The Jurassic dikes associated with the breakup of Pangea and the formation of the Atlantic Ocean cut through the rocks of the Agamenticus igneous complex, so its formation must have preceded the rifting event. But the Agamenticus igneous complex must be younger than the Acadian mountain-building event because it intrudes the Webhannet granite of early Devonian age. It is also younger than any igneous rocks of Carboniferous age. Thus it is difficult to explain the origin of this body of rock: it is too old to be related to the rifting of Pangea, too old to be related to the passage of the New England hot spot, and too young to be related to the Acadian mountain-building event.

Recent dating of one phase of this complex gives an age of 228 million years, placing its formation in the lower Triassic period. Watch for the almost continuous exposures of Agamenticus igneous rocks along the southern 3 miles of the Maine Turnpike north of the tollbooths. A few of these outcrops have vertical basalt dikes.

Layered Gabbro at Cape Neddick

The nearly circular mass of dark gabbro in the Cape Neddick pluton is perhaps the youngest rock in Maine; it is dated at 118 million years old, placing it near the middle of the Cretaceous period. The gabbro is layered as though it were a sedimentary rock, and, in a sense, it is.

If you look closely at the gabbro exposed near the parking area across from the Cape Neddick Light, you see thin layers that differ in their mineral composition. Some have more green olivine, others black pyroxene. In places, these layers lie one on top of the other like sheets of paper. In other places, they cut across one another, as though a current had eroded them. In a few places, you can see what looks like the outline of a channel that was eroded in the layers, then filled. All of those features are the sort of thing you normally expect to see in a sandstone, not in an igneous rock.

This kind of layering is actually fairly common in large masses of gabbro, and geologists debate endlessly about its origin. Most now

Vertical layers of light-colored minerals in the Cape Neddick gabbro. These rocks are exposed in the town park at Cape Neddick in York.

believe it forms as crystals of olivine, pyroxene, and feldspar grow in the molten magma and then settle through it to the floor of the magma chamber because they are denser than the molten magma. The rain of crystals onto the floor of the magma chamber does indeed resemble sand grains settling through water, and currents could flow through the magma chamber just as currents flow through water. So it seems reasonable that gabbro may in some respects resemble a sandstone.

You seldom see similar layering in granite because granite magma is too viscous to permit crystals to settle through it. The situation would be comparable to depositing sand from honey, instead of water. Basalt magma, which may crystallize into gabbro, is quite fluid, partly because it is extremely hot.

The outer zone of the Cape Neddick pluton is a breccia, a mass of broken rock solidly cemented together. It contains angular blocks of the enclosing Kittery formation set in a fine-grained matrix. The general aspect of the mess suggests violence, perhaps a steam explosion during emplacement of the gabbro.

The outer end of Cape Neddick is called the Nubble. A narrow gap eroded along fractures in the gabbro separates it from the main part of the cape. The Nubble Light was built in 1879 to help ships avoid running into what was then known as the Savage Rock.

Green, Black, and Pink Granite

Three very different colored granites of three distinct ages were quarried in southwestern Maine, all within a radius of about 4 or 5 miles. The Swenson Green granite quarry, which is in York, is a syenite, a rock like a granite but without grains of quartz. This rock is part of the Triassic Agamenticus complex. The quarry, located between U.S. 1 and I-95, provided foundation material for the Prudential complex in Boston.

The Minutti Black granite and the Spence and Coombs Black granite quarries are in South Berwick. Any igneous rock that will take a polish is considered to be a granite in the building trade. To geologists, this black rock is a gabbro of Cretaceous age and is part of the Tatnic Hill pluton. Polished sections of this rock sparkle with the

Nubble Light at Cape Neddick in York. The rock is the Cape Neddick gabbro, which is about 118 million years old and may be the youngest rock in Maine.

Swenson Pink granite quarry in Wells.

peacock iridescence of the feldspar labradorite. This stone has been used mostly for monuments.

The Swenson Pink granite quarry is on Quarry Road in the western part of Wells between Maine 4 and 109. The Swenson Pink is a true granite, part of the Webhannet pluton of Devonian age. The pink color occurs in the orthoclase feldspar from minute amounts of iron oxide within. Stone from this quarry was used in the Tomb of the Unknowns in the Arlington National Cemetery, Virginia.

Bald Head Cliff

Look for the brownish seams on the high cliffs of the Kittery formation just south of Ogunquit. Those are dikes of basalt that filled fractures during early Jurassic time, about 200 million years ago. That happened when the Atlantic Ocean was just beginning to open. Writer John McPhee calls these fractures "the stretch marks of the continent." Some of the basalt magma that moved through these fractures probably erupted as lava flows. Most of the dikes are parallel to the bedding in the Kittery formation. They range in width from a few inches to more than 10 feet.

A walkway from the parking lot of the Cliff House resort takes you to the top of the cliffs. Watch for the thin veins of white quartz

Thin beds of the Kittery formation intruded by thick basalt dikes at Bald Head Cliff in York.

that cut across the bedding in the Kittery formation, and for the basalt dikes that cut across them. The quartz veins probably formed while the rocks were folded, and the basalt dikes followed. Some dikes cut across other dikes.

If you look closely at the dikes, you will see that most of them are finely crystalline at their margins, a bit more coarsely crystalline within. When the molten magma squirted into the fracture, it quickly cooled against the Kittery formation, producing the fine margin. Then the interior of the dike crystallized more slowly, producing larger mineral grains.

Marginal Way

Marginal Way in Ogunquit is a public shore path that leads from Ogunquit north to Israel Head. The portion of this trail north of Perkins Cove has exposures of the same kinds of rocks found on Bald Head Cliff to the south—Kittery formation intruded by Mesozoic basaltic dikes.

Blowing Cave

A blow hole is a narrow opening that waves pass through, often to the accompaniment of little explosions as trapped air escapes. Blowing

Cave on Cape Arundel, near the north entrance to Kennebunkport Harbor, is one of the best blow holes in Maine. The sedimentary layers in the Kittery formation trend almost directly north, and a wide basalt dike cuts across them almost at a right angle. The waves eroded the dike to make a cave with a small hole in its roof. When the tide is rising, waves that break to the back of the cave shoot as much as 30 feet into the air.

Glacial Marine Deltas, Moraines, and Sand

Near the Kennebunk rest area and exit 19, I-95 crosses a broad plain. It is the far south end of the Great Sanford outwash plain, actually a series of glacial marine deltas. Nearby are larger deltas, the Merriland Ridge and Bragdon Road deltas, which resemble moraines. All contain a high proportion of sand.

Glacial marine deltas and DeGeer moraines near I-95 in Wells. —Modified from G. W. Smith, 1980

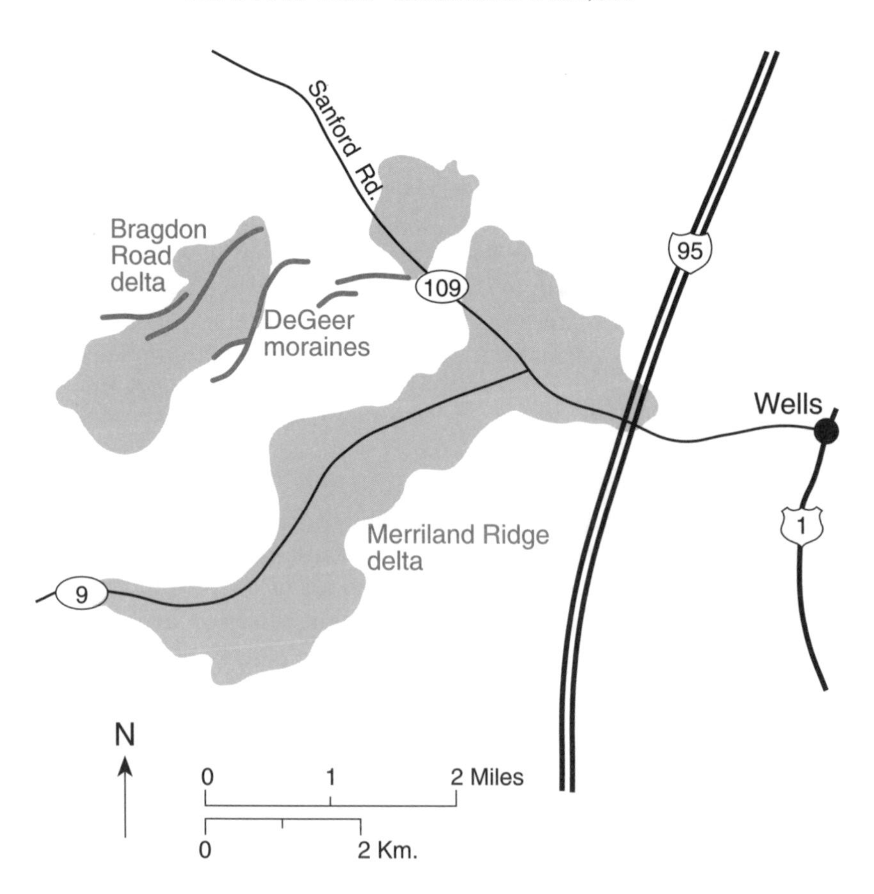

The area is so full of sand because it contains a great deal of granite, particularly in the Sebago pluton. Farther northeast along the coast, between Bath and Camden, granite plutons are few, and so are sandy outwash plains. Most granite crumbles when it weathers because the mineral grains do not interlock; the rock is like a jigsaw puzzle made of pieces with straight edges and square corners. Water soaks into the rock and reacts with grains of feldspar, converting them into clay. The feldspar grains swell as they change to clay, and that wedges the rock apart. The effect is about like what would happen if half the bricks in a wall were to swell.

The largest beaches in Maine are between Ogunquit and Portland, and at the mouth of the Kennebec River. Their sand came from at least three sources: erosion of glacial deposits along the Saco and Kennebec Rivers, as well as smaller streams such as the Scarborough,

Grand Beach and Pine Point in Scarborough.

Little, and Webhannet Rivers. Sand erodes from glacial deposits along the coast. Some sand may have moved landward from far offshore when sea level stood much lower, as recently as 11,000 years ago. For about 8,000 years sea level rose and waves pushed the sand to its present position.

All twenty or so sandy beaches between Kittery and Portland are between prominent bedrock headlands. Most are less than a half-mile long; Old Orchard Beach and Ferry Beach are the longest, more than 8 miles, with large tidal inlets at each end.

The jetties at several inlets were built to improve navigation, and in the belief that they would prevent tidal currents from moving sand into the harbors. The Wells Inlet jetty, built in 1962, trapped sand on both sides of the jetty as well as in the harbor. The accumulation of sand moved the shore some 600 feet seaward on either side of the jetty. Meanwhile, other beaches, deprived of that sand, have eroded, in both directions from the jetty. When the wind is right, the Wells Inlet jetty focuses the waves into the inlet, producing some of the best surfing along the Maine coast.

Excessive foot traffic partially ruined the sand dunes at Ogunquit Beach, and replacement sand was hauled in from nearby glacial deposits. Wave action wears the grains of normal beach sand smooth; the grains of glacial sand tend to have sharp edges because they are less worn. Barefoot bathers and beach strollers contend that the sand hurts their feet. Beach sand also contains nutrients required for the growth of dune grass, which glacial sand lacks. So the artificial dune made of glacial sand grows grass only near a sewage treatment outfall, and much of the rest is blowing away.

Bouncy Bridges and Shaky Glacial Mud

The Presumpscot formation is a glacial mud named for exposures on the banks of the Presumpscot River near Portland. These sediments are usually visible only in recently dug holes. Where you do see it, the mud is gray or bluish gray, and normally without prominent layers. Bleached white or yellow clam and snail shells abound, and they generally fall apart at the touch.

Along I-95 particularly, but anywhere that the mud of the Presumpscot formation is thick, you see a special kind of bridge construction. Watch the bridges over the turnpike near Portland, especially the one at exit 44, to I-295. Watch for abutments that begin more than 20 feet from the edge of the road, with a pile of dirt between them and the road. The pile of dirt is a spread footing intended to stabilize the mud. If the abutment starts to sink it displaces the

Spread footing at bridge abutment over I-95 in South Portland. The overpass is the southbound entrance at exit 44.

mud under it, which then rises around the abutment. The spread footing holds the mud under it down, and that holds the abutment up.

Another method of stabilizing the Presumpscot mud is to remove the water from it, making it strong enough to hold the weight of a bridge. Still another way to deal with it is to dig it up and replace it with gravel. But neither of those techniques is possible where the mud is hundreds of feet thick. In those places, spread footings are the only solution.

If you feel a little bump as you drive onto a bridge, it may mean that the highway engineers are having trouble with the glacial mud under it. If you have a smooth ride on and off a bridge in the Presumpscot region, the engineers won.

Mystery of the Scarborough Elephant

A number of years ago, a man digging an irrigation pond in the Presumpscot glacial marine mud near Scarborough found that he could not move the backhoe through stuff that normally has the consistency of warm butter. He discovered that the problem was a 5-foot-long elephant tusk. It was near the top of the Presumpscot formation, below about 5 feet of sand.

Further investigation established that the tusk belonged either to a modern elephant or to an ice-age mammoth; a tooth would determine which. The search for a tooth ended when a local reporter found

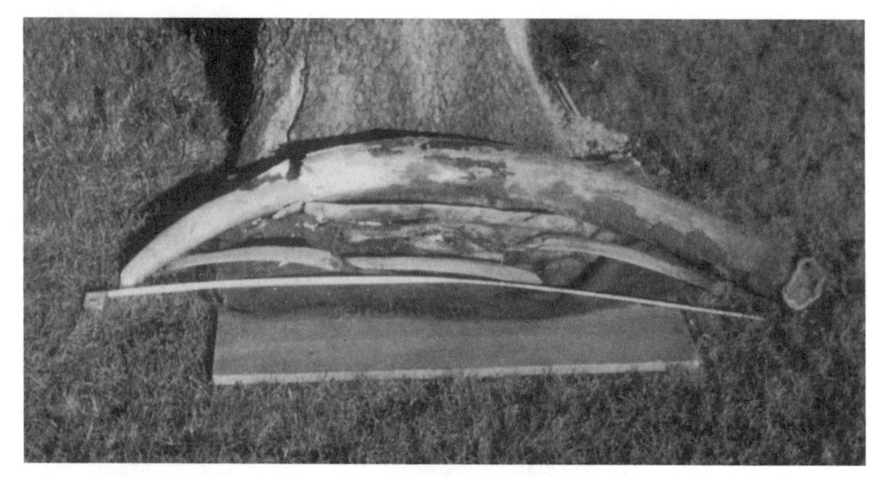

Mammoth tusk with portions of ribs found in Scarborough in 1959. The tusk was buried in the Presumpscot muds and has recently been dated at 10,500 years old. —Arthur Hussey photo

a story in an 1816 newspaper describing how an irate farmer shot a circus elephant near this spot. The murdered elephant, Old Bet, had been exhibited by Hackaliah Bailey.

In 1990, historical research revealed that the remains of Old Bet had been taken to New York for display, so the Scarborough tusk was not hers. The tusk, missing for a number of years, was found in the cellar of an eccentric collector and acquired by the Maine State Museum in Augusta, which obtained a radiocarbon date of 10,500 years. And it has now been positively identified as belonging to a mammoth, as teeth, another tusk, and some bones have since been found on the site.

The position of the tusk between the Presumpscot clays and the overlying marine sand, its radiocarbon date, and the elevation of the site at about 100 feet are hard to explain. Other studies suggest that at the time of the radiocarbon date, sea level was more than 100 feet below its present stand. And the sand and mud in which the Scarborough tusk was found were deposited about 13,500 years ago.

The summer Old Bet was shot was a bad one for Maine farmers, the worst in history. The year 1816 was remembered as "Eighteen Hundred and Froze to Death" and "the Year without a Summer." Each of the summer months brought snow and freezing temperatures to Maine. Crops failed. It now seems clear that volcanic ash erupting from Tambura, in Indonesia, in the fall of 1815 caused the cold weather.

The cloud of ash and volcanic gas spread through the atmosphere, blocking the heat of the sun, especially in the Northern Hemisphere. The eruption of Pinatubo in the Philippines in 1991 similarly cooled New England in 1992 and 1993, although not as severely.

U.S. 1
The Peninsula Coast
Portland—Ellsworth
141 miles

The rather straight southernmost coast of Maine trends generally northeast, parallel to the rock formations and to the faults that break them. Between Portland and Rockland, the coast consists of a series of long peninsulas, with embayments between them.

Casco Bay

The peninsulas, embayments, and islands in Casco Bay have a northeast trend like those farther south, except that the embayments are more abundant. That difference may reflect a greater abundance of faults in the Casco Bay area than farther south. Indeed, the largest embayment along the southern section of coast is Saco Bay, which was eroded along extensions of the Casco Bay faults. Between Harpswell Neck and Bailey and Orrs Islands the peninsulas become even more pronounced, but they still follow a northeast trend.

Tradition has it that Casco Bay contains 365 islands, one for each day of the year; thus the name Calendar Islands. It is hard to count the islands because so many of the smallest ones appear only at low tide. The actual number of islands in Casco Bay is probably less than two hundred, even counting the rocks that appear only at low tide.

The more northerly trend of the peninsulas east of Casco Bay reflects a change in the bedrock structure. From the air, the shoreline resembles the bones of a hand. The peninsulas, bays, and islands all owe their origin to differential erosion of alternating bands of weak and strong rocks. The bays follow faults and beds of weak sedimentary rocks. The peninsulas and islands are made of stronger stuff—granites, quartzites, and other resistant rocks. Ancient rivers eroded their valleys along the zones of weaker rock, then seawater flooded them. Glacial erosion probably played a minor role in shaping the coastline.

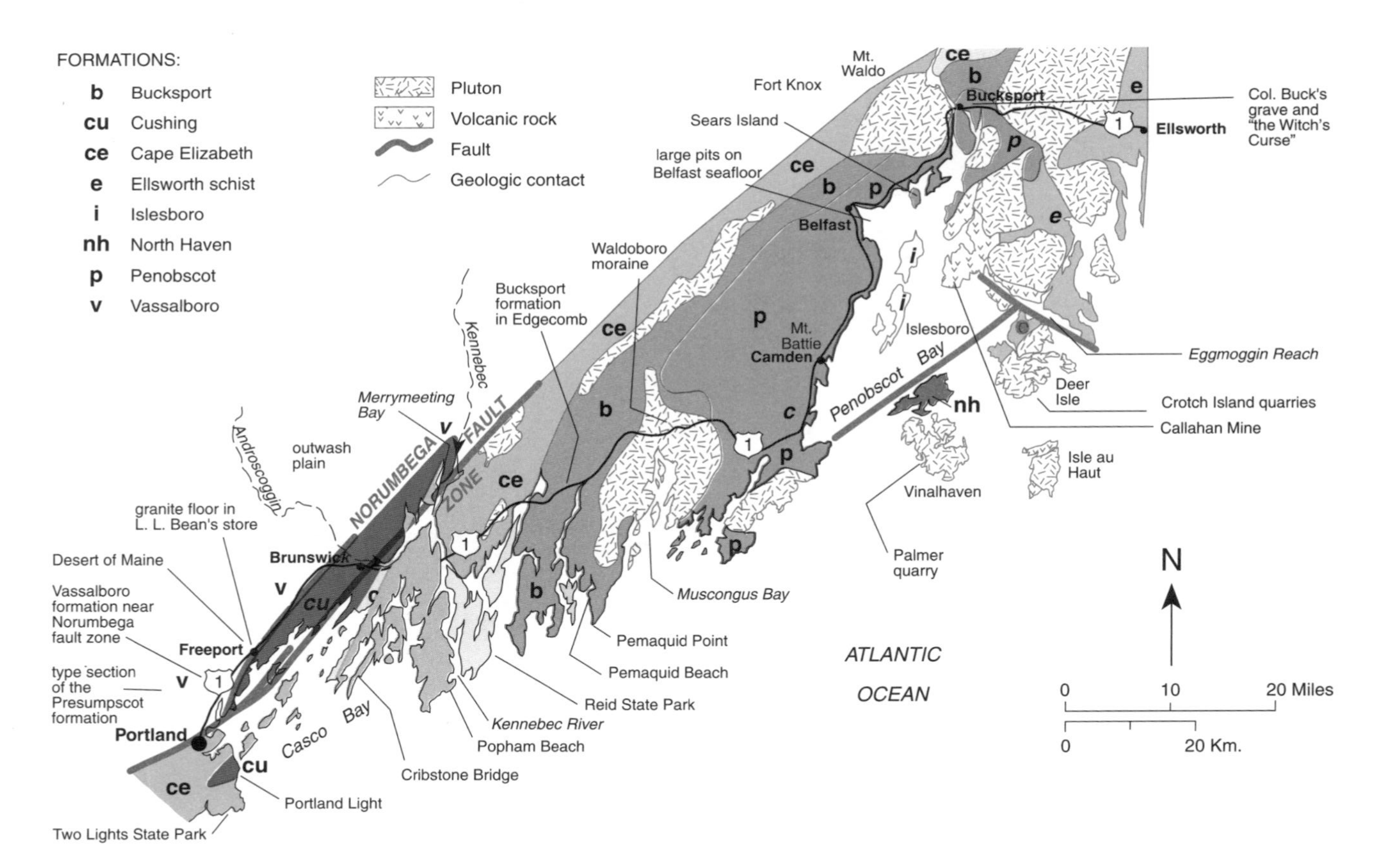

Geologic features along U.S. 1 between Portland and Ellsworth.

Elongate peninsulas, bays, and islands in Casco Bay. The bedrock here is the Cape Elizabeth formation.

This stretch of heavily embayed coast has long been used in elementary geology texts as an illustration of a drowned shoreline. Sea level has most certainly risen along this coast, but it has also risen on the whole Maine coast, along with the sea level in most of the world; only a few shores resemble this section of the Maine coast. The shape of this coast has a lot more to do with the bedrock structure and the alternating zones of weak and resistant rocks than with changes in sea level. The same rocks extend far inland as well as offshore, so the coast would look much the same whether sea level rose or fell.

Portland

The first settlement in this area was near the Portland Head lighthouse in 1675. It was called Falmouth. Indians soon destroyed the town and scattered the settlers. A second settlement begun in 1716 suffered no further Indian attacks, but the British burned much of the town in 1775. The town took the name of Portland in 1786. The state capital was in Portland when Maine became a state in 1820; it was moved to Augusta in 1832. Portland has long been the largest city in Maine.

Casco Bay Group

The Casco Bay group of metamorphic rock formations extends along the coast from near Saco to close to Bangor. The rocks were originally mudstones, limestones, sandstones, basalt, and rhyolite. The most widespread units are the Cape Elizabeth formation and the Cushing formation, which weathers to a rusty brown color. Erosion along a number of northeast-trending faults, as well as the weak rocks of the Cushing formation, opened Casco Bay.

Two Lights State Park

Two Lights State Park is in Cape Elizabeth, on the Cape Elizabeth formation. Most of the original sediments in the Cape Elizabeth formation were mudstones of uncertain age. The minor metamorphism did not erase the typical gradation from coarse sediment at the base of each layer to fine sediment at the top.

The Cape Elizabeth formation continues northeast well past Wiscasset. The numerous folds and faults at Two Lights State Park probably date from the Acadian mountain-building event of Devonian time. The gray schist in which the cleavage planes intersect folded bedding looks like the surface of a roughly sawed log. Samples of it and of similar rocks along the coast are often mistaken for petrified wood and submitted to the Maine Geological Survey and college geology departments for verification.

The park contains a small pocket beach between two bedrock headlands. It migrated shoreward with rising sea level, isolating a salt marsh from the ocean, converting it to a freshwater swamp.

Portland Head Light

President George Washington authorized construction of this famous lighthouse in 1791; it is the oldest light on the East Coast. It stands on the Cape Elizabeth formation and the rusty Cushing formation. The Portland Head Light is one of few not rebuilt since its original construction.

Vassalboro Formation and the Norumbega Fault Zone

The Norumbega fault zone separates the Avalon terrane from a wide belt of mudstones that were deposited in deep water during Silurian and Devonian time. As the name suggests, the fault zone is actually several more or less parallel faults.

Between Portland and Brunswick, two of those faults enclose a slice of the mudstones in the Vassalboro formation. Granite magma injected between layers of the mudstones converted the formation into

Cape Elizabeth formation at Two Lights State Park. The ridges under the hammer are mullion structure, accumulations of quartz causing the rock to swell there. This rock is commonly thought to be petrified wood. —F. J. Katz photo, U.S. Geological Survey

nearly horizontal sheets of dark schists and light granite. Rocks of this type probably form the basement under the Silurian slate belt northwest of the Norumbega fault. Similar rocks, but with less granite, crop out along I-295 as far as Brunswick.

Presumpscot Formation

Watch along the Presumpscot River near I-95 for exposures of glacial marine clays of the Presumpscot formation, the only glacial sediment in Maine that enjoys formation status. The gray to bluish gray deposit of silt and clay fills many low areas along most of the Maine coast, and inland along the large rivers. It reaches some 20 miles up the Saco River and more than 100 miles up the Penobscot River. Meltwater streams dumped the sediment into the shallow seawater along the ice front.

Much of the sediment is rock flour, pulverized rock scraped off outcrops and ground within the glacier. Most of the Presumpscot for

Dark schist and light bands of granite in the Vassalboro formation on I-295 in Falmouth.

mation contains fossils of clam and snail shells. Radiocarbon dates show that they lived between 13,500 and 12,500 years ago.

Clay from the Presumpscot formation has been used as a home remedy for bee stings and poison ivy, as surfacing for tennis courts, and as raw material for bricks. In the late 1800s, nearly one hundred brick plants in Maine produced nearly one hundred million bricks a year. Two plants survive, in Gorham and Danville. The wet mud is cast into the shape of bricks, dried for several days, and then baked at temperatures of more than 2,000 degrees Fahrenheit for several days.

L. L. Bean's Floor

We hope that the granite floor in the ground level of Bean's store in Freeport lasts a long time. The slabs of pink granite came from Deer Isle. They were cut from granite left over from the paving of a portion of a military airfield near Washington, D.C.

Look for the large blocky crystals of pink orthoclase feldspar with thin rims of white plagioclase feldspar. They are characteristic of a rare rock called rapakivi granite, after a town in Finland. Many geologists have traveled long distances to see outcrops of rapakivi granite. Here you can see it in nicely sawed slabs. Many of the slabs show

a high percentage of dark minerals, more than is normal in granite. In some cases almost half the rock is black. The Deer Isle granite and others along the eastern coast of Maine were intruded by basalt before they completely solidified. This added dark minerals to these granites.

The Desert of Maine

Many people associate dunes strictly with deserts, but you can find them in all sorts of places that are not deserts. The Desert of Maine is a commercialized example of a tract of sand dunes in a humid climate, one of several in Maine.

John Tuttle settled this land in 1797, pulling his house along behind twenty-four yokes of oxen. After trying crops, he soon turned to sheep. They crop the grass very close and finally pull up the roots. The sheep soon exposed the sand, and then the wind began to blow it.

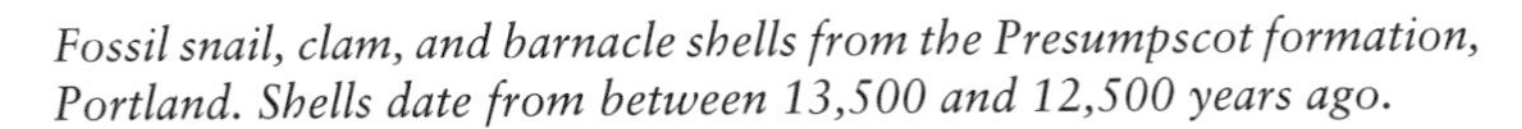

Fossil snail, clam, and barnacle shells from the Presumpscot formation, Portland. Shells date from between 13,500 and 12,500 years ago.

Deer Isle granite on the floor at L. L. Bean's, Freeport. Ladies' size 7B boot for scale. —Photo by permission

The sand is the upper part of the Presumpscot formation. Several feet of sand were deposited in shallow water as the sea withdrew, and it typically covers the glacial mud. In many places in coastal Maine, the wind blew this sand for a short time after glaciation, before plants covered and stabilized it. Anything the wind moved once, it can move again. If you expose this sand to the wind by destroying its plant cover, it will blow into dunes.

Between 1903 and 1936, the Desert of Maine increased from about 75 to more than 300 acres. Given a chance, plants would probably overgrow and stabilize the dunes. Even larger areas of windblown sand exist near Leeds and Wayne, and in the Kennebec River valley. All have a history similar to that of the Desert of Maine.

Bowdoin College–Brunswick Naval Air Station Sand Plain

Both Bowdoin College and the Brunswick Naval Air Station to the east are on a broad sand plain deposited while this part of Maine was submerged as the last ice age was coming to an end, around 13,000 years ago. The flat surfaces of glacial outwash and deltaic deposits easily accommodate airfields. And outwash deposits make easier digging than bouldery glacial till, so they are more commonly chosen as sites for cemeteries than are glacial moraines.

The well-drained and acidic soils of such deposits commonly support lush growths of white pine, such as the Bowdoin Pines on Maine 24 just east of the college. These pines are 90 feet high and about 125 years old. In colonial days, large, straight pines like these were marked with a broad *A* as the king's timber, to be used only as masts for the British navy. Cutting them for private use was a capital offense.

In Brunswick, not far from the Pines, is the Stowe House, now an inn. It was the home of Harriet Beecher Stowe and her husband while he taught at Bowdoin College and she wrote *Uncle Tom's Cabin.* Bowdoin College was founded in 1794 and named for James Bowdoin, then governor of Massachusetts and Maine. Henry Wadsworth Longfellow, Nathaniel Hawthorne, and Franklin Pierce were at Bowdoin together early in the 1820s.

Merrymeeting Bay

Merrymeeting Bay, a major stopover for migratory birds, is a body of open water at the junction of the second largest river in Maine, the Kennebec River, with the third largest, the Androscoggin. Four other small rivers also merge in this bay. Merrymeeting Bay was probably a lake when sea level was low during the ice ages.

The long direction of the bay is from northeast to southwest, parallel to the Norumbega fault zone, as well as to the bedding of the Cushing formation, which erodes easily. The outlet to the sea, the Kennebec River, leaves the bay midway along its southeastern shore and flows south along the trend of the rocks south of the Norumbega fault zone to Popham Beach.

Harpswell Neck and Orrs and Bailey Islands

Harpswell Neck next to Casco Bay and Orrs and Bailey Islands to the east are two of the longest and thinnest peninsulas on the Maine coast. Both are outcrops of the Cape Elizabeth formation. The intricate crenulations of the shoreline express the complicated folding of the rocks and the variations in their resistance to weathering and erosion.

Basin Cove Falls on Harpswell Point reverses with the tide in Casco Bay. The volume of Basin Cove is so great that water cannot flow through the narrow inlet fast enough to keep pace with the rise and fall of the tides. So water cascades in or out through the inlet with every turn of the tide. The current both ways is fast enough to interest white-water canoeists and kayakers. Like many similar tidal basins in Maine, Basin Cove was the site of a mill that used the strong tidal flow for power.

Cribstone Bridge, connecting Orrs and Bailey Islands in Harpswell. Granite blocks were quarried in Pownal.

Cribstone Bridge

Cribstone Bridge, also called the Cobwork Bridge, was built in 1928 to connect Orrs and Bailey Islands. It rests on blocks of granite quarried especially for this project in the Freeport-Pownal area. The long granite blocks with square ends were shipped from the quarries down the Cousins River to the construction site. The latticework of granite blocks that spans the 1,200 feet between islands allows tidal currents and sea ice to move freely through the bridge.

Except in the center span, the Cribstone Bridge snakes back and forth to stay on the good footing of the Cape Elizabeth formation. The center spans of 50 feet cross a deep channel through which lobster boats pass. No other bridge in the world has this type of foundation, although the designer is said to have gotten the idea from a bridge he saw in England.

Topsham-Phippsburg-Georgetown Pegmatites

Pegmatites are small bodies of granite composed of mineral grains several inches to several feet in size. They appear to form late in the crystallization of granite plutons, when the last residual magma becomes heavily charged with steam and commonly contains rare elements. Those concentrate in the residual magma because they do not fit well into the crystals of quartz, feldspar, and mica that make up ordinary granite. Some of these residual magmas finally crystallize within the mass of granite; some rise into the older rocks it intruded. Pegmatites typically form dikes, or irregular masses. Few are much more than a hundred or so feet across.

Mining of pegmatites in Maine began in the 1860s and continued until the 1960s. The main objectives were feldspar and mica. Feldspar was mined as an abrasive and a raw material in the manufacture of porcelain products. Large sheets of clear muscovite mica were first mined for isinglass for stoves, later for the insulating supports inside vacuum tubes. Some pegmatites were mined for gemstones, most notably aquamarine and tourmaline.

More abandoned feldspar quarries exist near the mouth of the Kennebec River, in Phippsburg on the west and Georgetown on the east. They are good places to look for gem beryl or aquamarine, as well as tourmaline in a variety of colors.

Popham and Reid State Parks and Beaches

The largest sandy beaches north of York and Cumberland Counties in southwestern Maine are at the mouths of the Kennebec and Sheepscot Rivers, in Phippsburg and Georgetown, respectively. Both beach systems are near the mouths of major estuaries, where fresh river water mixes with the salty tidal currents. Both beaches lie on the west sides of the estuaries, which have prominent bedrock headlands on their east sides.

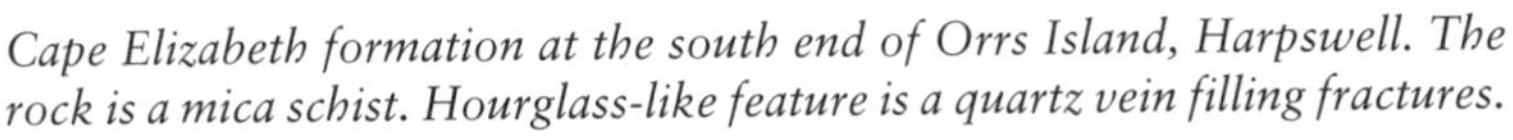

Cape Elizabeth formation at the south end of Orrs Island, Harpswell. The rock is a mica schist. Hourglass-like feature is a quartz vein filling fractures.

Waves move the sand to the west sides of the estuaries during heaving easterly storms. The rivers bring some of the sand to the coast from inland glacial deposits. The rest is from deltas deposited offshore during a low stand of sea level, then washed inland by waves as sea level rose.

Sandbars that emerge at low tide periodically connect the mainland to Wood and Fox Islands offshore from Popham Beach. When the Wood Island bar is attached to Popham Beach, it blocks eastward movement of sand along the beach. The accumulating sand builds Popham Beach as much as 300 feet seaward. When the Wood Island bar is eroded away, waves erode Popham Beach. On some occasions, the beach erosion was deep enough to destroy a number of summer cottages.

Wiscasset and the Sheepscot River

The Sheepscot River flows along a fault that juxtaposes the Cape Elizabeth formation, on the west bank, against the Bucksport formation. The bedding in these late Precambrian to early Paleozoic rocks is parallel to the fault and the river.

At the town of Sheepscot, about 4 miles upriver from Wiscasset, is a famous reversing tidal falls known for its Atlantic salmon fishing. The inland limit of tidal waters is 3 miles above the reversing falls, at Head Tide. An abandoned mill at this site used the Sheepscot River for power. Upstream from Head Tide, the Sheepscot River follows one of the several branches of the Norumbega fault.

Cottages collapsed in 1978 from wave erosion on Popham Beach, Phippsburg. Since this event, the beach has built out more than 100 yards. —Duncan M. FitzGerald photo

Bucksport Formation

The metamorphic rock between Wiscasset and Waldoboro is the Bucksport formation, a lower Paleozoic calcareous sandstone. In Edgecomb the Bucksport formation has vertical beds intruded with sills of granite. In Newcastle, on the Damariscotta River, the Bucksport formation is complexly folded.

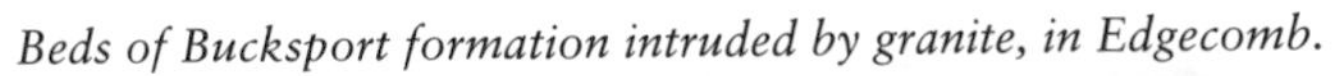

Beds of Bucksport formation intruded by granite, in Edgecomb.

Complex folds in Bucksport formation in Newcastle. At the tops of anticlines and the bottoms of synclines are folded beds that are refolded by the larger structures. —F. J. Katz photo, U.S. Geological Survey

Damariscotta

South of U.S. 1, the Damariscotta River is a broad tidal estuary in the outcrop area of the Bucksport formation. Between Newcastle and Damariscotta, the river is quite narrow, with reversing tidal falls at both the U.S. 1 and the U.S. 1A bridges. The river broadens above U.S. 1 into Salt Bay, a shallow tidal flat that attracts numerous shorebirds. The dam at Damariscotta Mills impounds Damariscotta Lake, which is 12 miles long.

Pemaquid Point

The bedrock at Pemaquid Point is the Bucksport formation, originally deposited as calcareous mudstone and sandstone, and here metamorphosed to a schist flaky with glistening crystals of mica. Its weathered appearance, like that of the Cape Elizabeth formation near Portland, resembles the grain of wood.

The arch of an elongate anticline is exposed near the center of the point. Just east of that is the trough of a syncline. Dikes and sills of very coarsely crystalline pegmatite granite intrude the schist, the dikes cutting across the layers, and the sills following them. One prominent sill makes a high ridge on the east side of the point.

Anticline in Bucksport formation at Pemaquid Point, Bristol. The light rock on the upper right is a portion of pegmatite granite sill. Pemaquid Light was built in 1827, marking the west entrance to Muscongus Bay.

Waves have deeply eroded the weaker layers in the formation, with the pegmatite generally being more resistant. Points like this one focus the energy of the waves on themselves, much as a lens focuses light. During heavy storms, the waves crashing on the point are quite obviously much larger than those that break along straighter sections of the coast.

A small museum in the old house formerly used by the keeper of the Pemaquid Light displays items used by Maine fishermen.

Pemaquid Beach

Pemaquid Beach and town park are just west of New Harbor, on the Pemaquid Peninsula. The beach lies between two outcrops of a quartz and biotite schist that has no formation name but resembles the Cape Elizabeth formation down the coast. Debris eroded from it accounts for much of the material on the beach. The quartz sand is bone white; the flakes of black biotite mica emphasize wave ripples and other little details.

Extremely low tides expose tree stumps and freshwater peat. The stumps are all that remain of a forest that grew in swampy ground landward of the beach a few thousand years ago. The sea level rose, and storm waves drove the beach landward until the stumps and the peat are now seaward of the beach. The smaller waves of summer transport sand onshore, often burying the stumps.

New Harbor, Round Pond, and Muscongus Harbor

The east shore of Pemaquid Point is quite straight, except in three places where natural harbors occur. New Harbor and Round Pond are in the Bucksport formation. Muscongus Harbor is formed in Devonian granite of the Waldoboro pluton. All harbors along this coast experience a tidal range of more than 15 feet. Eastern Egg Rock has the most southerly puffin establishment in Maine; it is also the closest to the coast. During the summer, you can take a boat from New Harbor to view the puffins.

Waldoboro Moraine

Waldoboro was founded by Samuel Waldo, who was part of the militia that in 1745 defeated the French and captured the great fort at Louisburg in Cape Breton, Nova Scotia. He was given the Waldo Patent, the area east of the Kennebec River and south of what is now U.S. 2, as a reward for his service. He brought a number of German immigrants to settle the area.

One of the largest and most continuous glacial moraines in the state lies just north of U.S. 1, in Waldoboro. Like most moraines in

Wharf at low tide in New Harbor, Bristol. The tide range along this stretch of coast is about 15 feet.

coastal regions of Maine, this one is a line of deltas deposited where the last ice sheet was grounded on the seafloor. Most of the material is sand and gravel, so gravel pits provide numerous nice exposures. They generally reveal minor amounts of till and Presumpscot clay, as well as sand and gravel.

Thomaston and Union Limestone

The largest active mineral industry in Maine, except for sand and gravel, produces Portland cement. The main raw material is marble, metamorphosed limestone of the Megunticook formation, just west of Rockland. Two deep quarries on either side of U.S. 1 in Thomaston mine the marble, which is then crushed, mixed with sand and clay, and roasted to make cement. The formula for many cements is based on the composition of a natural cement rock in England called the Portland Stone, which explains why brand names of cement commonly include the word "Portland." Cement is mixed with sand, gravel, and water to make concrete, a sort of artificial conglomerate.

For a number of years the Portland Dragon limestone quarry and the mill have been owned by the Penobscot Indian Nation, one of two groups of Native Americans living in Maine. A road leads south from Thomaston to Port Clyde where you can ride the *Laura B,* a mailboat to Monhegan Island. The village there resembles a nineteenth-century fishing village. The whole island is composed of Devonian gabbro

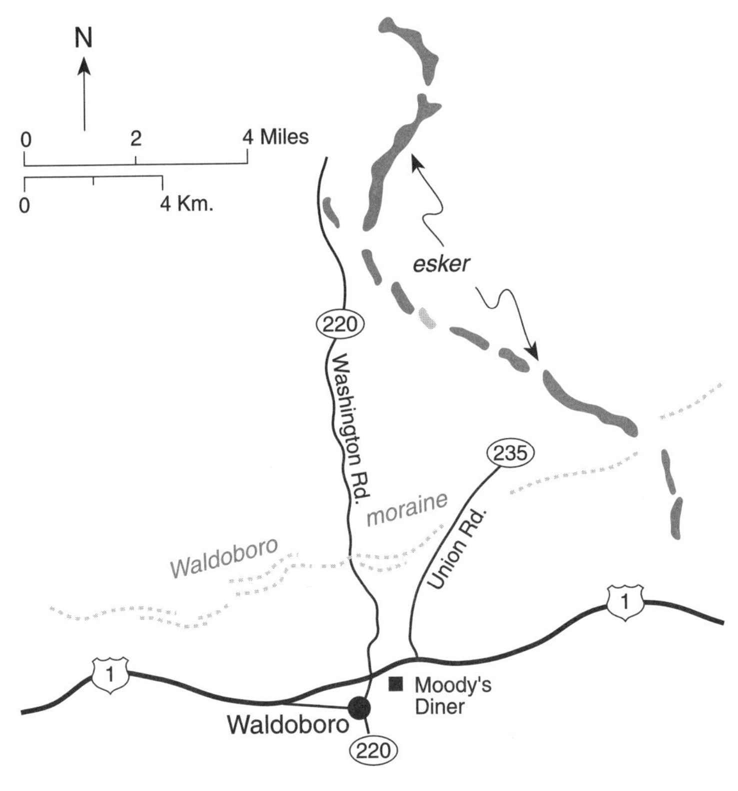

Waldoboro moraine and accompanying esker. —Modified from G. W. Smith, 1989

and the high black cliffs on the east side are at once foreboding and beautiful.

Penobscot Bay and the Bays of Maine Igneous Complex

U.S. 1 circles the west and north margins of Penobscot Bay. The western shore is approximately parallel with the strike of the bedding in the metamorphosed sedimentary rocks in this area.

Much of the bedrock on the many islands within Penobscot Bay, as well as on the eastern shore, consists of granite plutons intruded by basalt and gabbro. Similar rocks extend in Washington County as far as Calais. They are the Bays of Maine igneous complex. In many of these plutons, molten basalt intruded molten granite, forming basalt pillows or dikes in the granite. All of the igneous intrusions were emplaced during Devonian time, while the Avalon terrane was landing against America during the Acadian mountain-building event.

Many famous granite quarries operated on the islands, especially Deer Isle and Vinalhaven, with Penobscot Bay offering easy access to shipping. Granite quarries still work on Deer Isle. A number of mines producing such metals as copper, silver, and zinc have operated on the east side of the bay, from Cape Rosier to Blue Hill. Most recently, the Black Hawk Mining Company produced copper ore in the 1960s, and the Cape Rosier mine was working in the early 1970s.

Spruce Head Granite

The town of St. George lies on a peninsula at the southwestern end of Penobscot Bay. Most of the peninsula and numerous islands to the east are formed of the Spruce Head granite of Devonian age. The southern tip of the point is underlain by an unnamed Ordovician mudstone. These different rock types are separated by a thin sliver of the Benner Hill formation of Ordovician age along which Tenants Harbor has formed. The Hocking and Clark Island granite quarries worked the Spruce Head granite. Rock from the Hocking quarry surrounds a fountain and forms curbing at the United Nations building in New York City.

North Haven and Vinalhaven

Vinalhaven and North Haven are islands in the middle of the southern end of Penobscot Bay, both reached by ferry from Rockland. Bedrock on North Haven is a group of volcanic rocks, mostly basalt. The straight northern shore was eroded along the Owls Head–Deer Isle fault, which crosses the bay from Owls Head near Rockland in a straight line to Sedgwick. Bedrock on Vinalhaven Island is mainly granite that arrived as molten magma during Devonian time. It was quarried on the south end of the main island and on smaller islands offshore, including Hurricane Island. Most of the granite is light gray; some is pink. The Bodwell quarry produced the stone known to quarrymen and architects as black granite and to geologists as gabbro—the coarsely crystalline version of basalt.

The Vinalhaven Historical Society Museum has a fine exhibit of the granite quarries, including a galamander, a device with huge wheels for lifting and moving large blocks of granite. By tradition, all galamanders were painted light blue.

The Palmer quarry on the west side of Vinalhaven near the Basin was the source of some imposing pieces of granite. The eight solid granite columns, each 6 feet in diameter and more than 50 feet long, that surround the high altar in the Cathedral of St. John the Divine in New York City came from this quarry. The quarry blocks were to

Columns of granite from the Palmer quarry, Vinalhaven, being prepared in 1900 for the Cathedral of St. John the Divine in New York City. —Maine Historic Preservation Commission photo

have been turned into round columns on a special lathe, but they were too heavy, about 130 tons. So they were each cut in two pieces—one about 40 feet long and the other 15 feet long—and turned separately. The finished column pieces were shipped in 1903 on a barge, then loaded onto a wagon with solid wooden wheels, and pulled to the site of the cathedral by a steam-driven winch mounted on a tractor. The wagon broke every manhole cover it crossed. The eight columns were then set in place, and the church was built around them. This is the largest Gothic-style cathedral in the world.

Islesboro

Islesboro is in the middle of Penobscot Bay, connected to the rest of the state by a ferry from Lincolnville. The limestone and marble in the island are similar to those mined for cement at Thomaston and Union to the southwest, although they are considered a separate unit named the Islesboro formation.

A 647-million-year-old pegmatite intrudes marbles and schists of the Seven Hundred Acre Island formation, which must be older than the pegmatite. This pegmatite is believed to be the oldest accurately dated rock in Maine.

The Camden Hills, Mt. Battie, and Mt. Megunticook

The Camden Hills are higher than any others along the Maine coast, except those on Mount Desert Island. They dominate the western skyline as seen from Penobscot Bay. The Precambrian Islesboro formation, including its limestone member, underlies the area from Camden to Rockport. Camden Harbor was eroded along a fault that trends east and separates the Islesboro formation on the south from the Megunticook formation on the north.

The Megunticook formation consists mostly of schist that was baked in the heat of several nearby plutons. The original muddy sediments were deposited during Cambrian and early Ordovician time.

A gravel conglomerate that belongs to the Megunticook group of formations is the bedrock of Mt. Battie, which looms over the north side of Camden. It was deposited during early Cambrian time. The Megunticook formation is named for the outcrops of schist on Mt. Megunticook.

To the north and west, the Penobscot formation lies on the Megunticook group of formations. You can also see its rusty outcrops to the northeast, around Penobscot Bay through Belfast, Searsport, and Stockton Springs to the Penobscot River at Bucksport. Rocks similar to those in the Camden Hills occur near Calais, where they are called the Cookson group of formations.

Giant Pits in Belfast Harbor

At the northwest corner of Penobscot Bay lies the port of Belfast. A small river with a very long name, Passagassawakeag, drains into the harbor. This Algonquin name refers to the presence of sturgeon, but it is also a bad pun that describes what is happening in the harbor. Buried organic matter is decaying without the benefit of oxygen, thus generating both methane and hydrogen sulfide gases. This gas builds up in large pockets, then apparently bubbles violently through the mud, forming visible and smellable boils at the water surface and creating deep pits that resemble bomb craters on the harbor bottom.

Marine geologists from the University of Maine and the Maine Geological Survey discovered these pits while mapping the seafloor with a device called a sidescan sonar. They later recovered core samples from the bottom and actually visited some pits in the small research submarine *Delta*. The pits range upward to about 1,000 feet in diameter and 100 feet deep, although most are less than half that size. As many as 100 per square mile exist in the Belfast harbor seafloor.

The organic matter giving rise to the gas generation is not known, but it may be freshwater peat bogs buried with mud during the rise of sea level in the last 10,000 years. Lumber mills may have dumped into the harbor sawdust that was then buried by mud suspended during tide changes, and that could be the source of the gas. Natural gas deposits are formed in a similar way, but they are usually trapped in the earth and discovered through deep drilling.

The Strained Mt. Waldo Granite

The Mt. Waldo granite pluton is on the west side of the Penobscot River. A number of quarries worked this granite to supply building blocks for Fort Knox.

The rock in the Mt. Waldo pluton has the unusual property of deforming into drill holes and quarry faces, but blocks of the granite are perfectly stable when removed from the earth. Circular drill holes become oval in shape and rock explodes from quarry faces without warning. The one explanation for this curious behavior involves stresses being set up in the rock by surface temperature changes. If this theory is correct, you would think that all the granites in Maine would experience similar temperature changes and behave like the Mt. Waldo rock, but they do not.

Monument of Mt. Waldo granite, Frankfort, marking site where the rock was shipped.

Verona Island Bridge and the Penobscot Formation

On both sides of the river near the head of Penobscot Bay are towering cliffs that follow the vertical bedding in the Penobscot formation. The cliffs became the site of Fort Knox, and later required the construction of the highest bridge in Maine. It spans the gorge between Prospect and Verona, 137 feet above the Penobscot River. The bridge was completed in 1931 and is one of the most handsome in the state. Both ends are founded in the Penobscot formation, which you can see in the large, rusty outcrops at the western end of the bridge.

Bucksport and Colonel Buck's Gravestone

Colonel Jonathan Buck was living in Haverhill, Massachusetts, when the governor ordered him to execute a woman condemned as a witch. He later moved to Bucksport, where he is buried in the cemetery beside U.S. 1 and Maine 3. Legend has it that the likeness of part of a leg and a foot mysteriously appeared on his gravestone. Could it be the spectral image of the leg of the poor woman Colonel Buck executed? It is, in fact, a biotite inclusion altered from some piece of rock that fell into the magma chamber of the granite. A number of quarries in the Mt. Waldo granite just west of the Penobscot River produced rock full of biotite inclusions. One of them was undoubtedly the source of Colonel Buck's grave marker, as well as Fort Knox across the river.

Image of a human leg, known locally as "the Witch's Curse," on Colonel Jonathan Buck's gravestone at Bucksport.

Callahan mine in 1950, Cape Rosier in Brooksville.
—Fred Beck photo

Bagaduce Falls

Bagaduce Falls between Sedgwick and North Brooksville reverses its flow as the tides rise and fall. At low tide the water flows north across the falls, down the Bagaduce River, and into Bagaduce Bay. At high tide the water in Bagaduce Bay flows south across the falls. Flow speeds either way are great enough to inspire kayaking and white-water canoeing.

Blue Hill Ore Deposits

During the 1880s, the Blue Hill peninsula experienced a mining boom in copper and other metals, or at least a boom in mining stocks. The names of some of these mines have a distinctly western flavor: Star of the East, Excelsior, Eureka, Mammoth, and Bisbee do not sound like good Maine names. Most deposits had good shows of copper ore minerals but were too small to return a profit. The boom undoubtedly produced more money for the sellers of stocks than the buyers. A few mines operated off and on between the 1880s and the early 1970s, most recently the Callahan and Black Hawk mines.

The ore is in the Ellsworth schist and the Cranberry Island formation, near intrusions of Devonian granite. The ore minerals are metal sulfides, commonly in association with the worthless iron sulfide mineral pyrite. Exposure to water and air quickly converts pyrite to iron oxide minerals—rust. Its distinctive color flags the mineralized outcrops.

Deer Isle

The high bridge onto Little Deer Isle crosses Eggmoggin Reach, a wide and straight connection between the waters of the upper Penobscot Bay and Blue Hill Bay to the southeast. It follows a fault. A river probably eroded a valley along the fault, then the glaciers gouged it out into a broad trough. Another major fault extends across the southern end of Penobscot Bay from Owls Head to the Eggmoggin Reach fault. It separates the rocks of Deer Isle from those of Little Deer Isle.

The famous Deer Isle granite quarries are in the southern part of the island, in Stonington. The largest are on Crotch Island in Stonington Harbor. They produce an unusual pink granite. Pink granite in Maine is generally restricted to the Avalon terrane, having crystallized during Devonian time. No one that I know of has explained why. This granite contains large crystals of feldspar that are pink in the middle and thinly rimmed in white. Without going into detail, I can say that the pink interior is orthoclase feldspar, the white rim plagioclase feldspar. This peculiar kind of rock is called rapakivi granite, after a place in Finland.

Rapakivi granites are quite rare, every bit as baffling as they are beautiful, and the subject of a voluminous literature. Many geologists argue that the pink orthoclase feldspars began to crystallize in one place in the magma chamber, then acquired their white rims as they moved to another place where higher temperature favored plagioclase. This rapakivi granite also contains more dark minerals rich in iron than you normally see in granite. That may be related to the intrusions of mixed granite and basalt from Penobscot Bay eastward along the coast of Maine.

This granite is part of the John F. Kennedy Memorial in the Arlington National Cemetery in Washington. And you can see it in the slabs that cover the first floor of the L. L. Bean store in Freeport.

Blue Hill and Blue Hill Falls

The town of Blue Hill was the center of a mining boom in the middle of the 1800s. Properties were bought and sold, stocks issued, and even a little mining was done. Only the mines on the west side of the Blue Hill peninsula, on Cape Rosier, continued to operate after this boom, finally closing in 1972.

Blue Hill towers over other elevations on the east side of Penobscot Bay and is visible from all directions. It is Ellsworth formation at its contact with the Lucerne granite. The heat from the magma baked and toughened the rock enough that it better resists weathering than others nearby, and so it stands high.

Blue Hill Falls is one of the largest of the reversing tidal falls in Maine. Tidal waters are slowed by the narrow opening between Blue Hill Neck and the mainland, allowing the high tide to build in Blue Hill Harbor while the water in Salt Pond sits at low tide. The difference forces the water in the pond with such velocity that high standing waves are formed, delighting the canoeist, the kayaker, and inner-tuber alike. When the tide falls in Blue Hill Harbor you can ride out on a similar current.

Lucerne Granite

Between Bucksport and Ellsworth, U.S. 1 crosses hilly terrain eroded on the Lucerne granite. The granite is white, with large feldspar crystals set within dark halos of crystals of biotite, hornblende, and other black minerals. Blueberry barrens develop in the sandy and bouldery glacial deposits that cover the Lucerne granite.

Down East Maine: Bangor and Ellsworth to Calais

The fastest way to sail is downwind. Along the Maine coast the wind often comes from the southwest, so downwind takes you eastward up the coast. In the rest of the country, *Down East* refers to the whole state of Maine, while in Maine the term means the easternmost coastal section of the state. The farthest Down East you can get in Maine is Hancock and Washington Counties. Washington County has signs advertising it as the Sunrise County, claiming that the sun rises here first. Perhaps, but Cadillac Mountain on Mount Desert Island, Katahdin, and Mars Hill all have an earlier sunrise at some times of the year.

Almost half of the bedrock of Hancock and Washington Counties is plutonic igneous rock, mostly granite, with smaller areas of gabbro. All of it formed during Devonian time. Bedrock in a large area between Machias and Calais is volcanic rocks, rhyolite, and basalt that erupted during Silurian time. All of these igneous rocks are probably aspects of the Acadian mountain-building event, the collision of the Avalon terrane with North America. Bedrock elsewhere is small patches of sedimentary and volcanic rock that range in age from latest Precambrian to middle Devonian.

Bedrock geology of Down East Maine and Mount Desert Island.

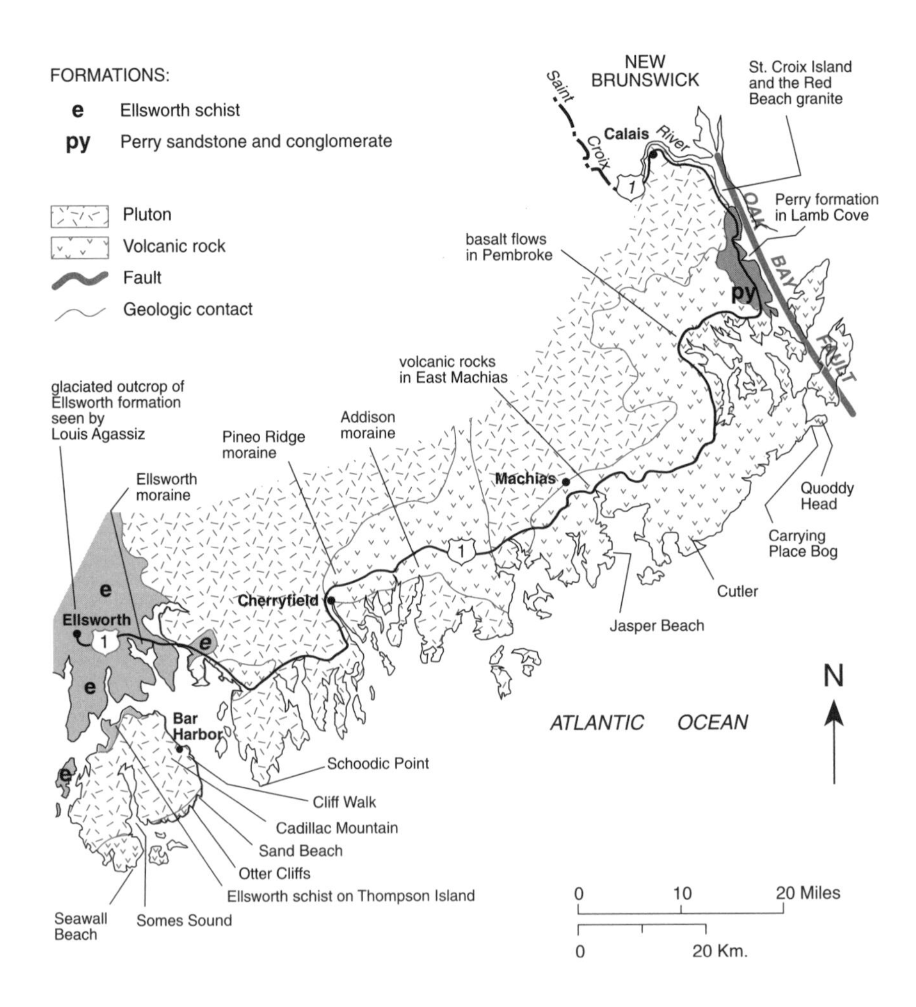

Eastern Maine has some spectacular glacial moraines and deltas deposited when seawater flooded across the land. U.S. 1 follows the crests of several of these long moraines between Ellsworth and Machias.

The Moving Continental Crust

Cliffs of volcanic rocks line the coast of Washington and Hancock Counties; many are between 50 and 150 feet high. Most of Maine has been rising since the glacier melted at the end of the last ice age, removing a heavy burden from the floating continental crust, but the coastal area of Washington County is sinking. Survey markers, or benchmarks, installed in eastern Maine in 1916 were resurveyed in 1978. Those in Washington County had gone down, relative to benchmarks in central Maine. Submerged roads, docks, dikes, and ship-launching sites provide other evidence for the subsidence of eastern Maine. Meanwhile, central Maine has seen little recent change in elevation, and northern Maine is slowly rising. No one knows why eastern Washington County is sinking.

The resurvey of 1978 also revealed changes in the horizontal distances between benchmarks. The distance between a point in Maine and one 17 miles away in New Brunswick was measured with an extremely precise laser surveying instrument in 1975, and again in 1983. In that time, the distance shortened by a little more than 6 inches, a phenomenally fast movement. Little wonder that the region feels a number of small earthquakes every year.

Bays of Maine Igneous Complex

The term *Bays of Maine* generally refers to the embayed coast between Penobscot Bay and Calais. A characteristic but most unusual feature of the igneous rocks in the Bays of Maine is the occurrence, often in the same outcrop, of granite and basalt or gabbro. The Silurian volcanic rocks between Machias and Calais have a similar range in composition.

Geologists ordinarily think of granite and basalt as rock types so distinct that they are virtually incompatible. Granite magma forms through melting of the continental crust, basalt magma through melting of the upper mantle. In the Bays of Maine igneous complex, they occur together, in associations that can only be described as intimate.

One of those is the occurrence of blobs of basalt in granite. The blobs look like the basalt pillows that form where a lava flow erupts underwater. Many geologists contend that they formed when basalt magma, with a melting temperature of about 2,000 degrees Fahrenheit, intruded granite magma, with a melting temperature of about 1,200 degrees Fahrenheit. Both magmas were hot, but the basalt magma was the much hotter of the two. It quenched against the molten granite almost as though it were erupting into cold seawater. In the Bays of Maine complex, the basalt pillows formed during Devonian time, while the Acadian mountain-building event was underway.

Maine 3, Maine 102, and the Park Loop Road
Mount Desert Island and Acadia National Park

About 100 miles

Maine 3 extends from Ellsworth to the north and eastern half of the island. Maine 102 gives access to the western portion of the island. The Park Loop Road, which begins at the park headquarters on Maine 3 near Bar Harbor, reaches many of the most popular sites within the southeastern portion of the park. Much of that road is one-way, clockwise.

A Bit of History

Samuel de Champlain and his band saw Mount Desert Island as they sailed the Maine coast in the fall of 1604. The summits of its mountains were as barren then as they are now, and Champlain named it *Isle des Monts-Deserts,* island of the deserted mountains. And that reminds us how to say this often mispronounced name: while spelled like the dry place, it should properly be pronounced like the after-dinner sweet, although both pronunciations are common and acceptable in Maine. The word *desert* as originally used in English came from the French and meant barren. It was not until the nineteenth century, when Americans entered the arid West, that *desert* came to mean a dry region.

In 1688, King Louis XIV of France gave the island to Sieur de la Mothe Cadillac, but the English forced Mothe to leave in 1713 and gave the island to the British governor of Massachusetts, Sir Francis Bernard. Cadillac went west and founded Detroit. After the Revolution, Cadillac's granddaughter and Bernard's son managed to gain control of separate halves of the island. Green Mountain, the highest on the island, became Cadillac Mountain.

Of the several versions of the origin of the word *Acadia,* it seems most probable that it is the shortened form of the name of the Shubenacadie River that flows into the eastern end of the Bay of Fundy in Nova Scotia. The French called this region *l'Acadie.* The English changed this to Acadia.

The French lost control of Canada and Maine during a number of wars with the British and American colonists in the middle of the eighteenth century. In 1745, the great fort at Louisburg in Cape Breton, Nova Scotia, was captured by an army that consisted largely of Maine citizen soldiers. The Acadians in Nova Scotia refused to swear allegiance to the King of England and were exiled to various American colonies. A number of them settled along the upper St. John River in Maine and New Brunswick, where they became successful loggers and farmers. Many went to Louisiana, where the pronunciation of their name became "Cajun."

Acadia National Park, which includes about 75 percent of Mount Desert Island, is the nation's smallest and second most visited national park. Maine originally set aside some of the island in 1901 as a public reservation. In 1916, the land was transferred to the federal government, which established the Sieur de Monts National Monument. The name was changed to Lafayette National Park in 1919, for the French general who assisted the colonies in the Revolutionary War; it was finally changed to Acadia National Park in 1929.

By the middle of the nineteenth century, a number of visitors began to spend their summer vacations on Mount Desert Island. At first, they camped in the woods and by the shore, fishing and hunting. In the 1870s and 1880s, a more elegant sort of visitor arrived, and grand hotels were built in Bar Harbor and elsewhere on the island. They came by train and steamship, many directly from Boston and New York in their private rail cars. Perhaps because the fashionable ground in Newport, Rhode Island, was already claimed toward the end of the 1880s, a few wealthy visitors built elaborate cottages, and high society moved from the towns to more remote parts of the island. Some of these cottages still stand, although most now have other uses. Many were destroyed in the great forest fires that beset many parts of Maine, including Mount Desert Island, during the summer and fall of 1947.

Geology of Mount Desert Island

Granites of Devonian age dominate. The generally circular outline of the island expresses the shape of the granite plutons in the bedrock. The Cadillac granite, named for the highest mountain on the island, underlies most of Mount Desert Island. The oldest rock is the Ellsworth schist of early Paleozoic age; it appears in the northern and western parts of the island. Near the southern end are a series of Silurian volcanic rocks, mostly rhyolite, called the Cranberry Island series; they belong to the coastal volcanic belt. Devonian granites are the youngest bedrock. Ice-age glaciers eroded great valleys and left thin deposits of glacial debris in the low areas. A number of sand and pebble beaches contain material derived from both the bedrock and the glacial deposits.

Ellsworth Schist

The bedrock between Ellsworth and Mount Desert Island is the Ellsworth formation, metamorphic rocks of uncertain age. They had already endured a complex history before Silurian time began, when they were exposed, eroded, and then covered by Silurian sedimentary and volcanic rocks. So the Ellsworth formation must be older than the Silurian rocks that cover it. It was originally deposited as mudstones, partly interbedded with thin volcanic rocks. They were complexly folded, probably during the Acadian collision of Devonian time, although they appear to have had an earlier history as part of Avalonia before they arrived in North America. In fact, some evidence suggests that the Ellsworth formation was metamorphosed into schist before the Acadian mountain-building event.

Good exposures of Ellsworth schist exist on the north end of Mount Desert Island, particularly near the picnic area opposite the park in-

Weathered outcrop of Ellsworth schist on Thompson Island, Trenton.

formation center on Thompson Island. Watch for them as you cross from Trenton. As usual in schists, the minerals are segregated into bands, in this case pale bands of quartz and feldspar, and darker bands of the greenish mica chlorite. The mica flakes are aligned more or less parallel to each other, giving the rock its tendency to break into slabs.

An outcrop of the Ellsworth formation played an important role in the development of geologic thought in the United States. Louis Agassiz, a Swiss scientist, showed that glaciers had once covered much of Europe. In 1849 he was appointed professor of natural history at Harvard University. On a trip to Bar Harbor in 1857, Agassiz examined an outcrop of the Ellsworth formation in Ellsworth Falls and argued that its smooth surface with parallel scratches was evidence that glaciers had covered Maine. The outcrop is southwest of U.S. 1A, on the grounds of Foreign Motor Car. It probably looks much as it did when Agassiz saw it. Others had arrived at the same conclusion several years earlier, but Agassiz was the better publicist.

Glaciated outcrop of Ellsworth schist in Ellsworth Falls. Upon seeing this outcrop in 1857, Louis Agassiz proclaimed that Maine had been visited by the great ice sheets.

Bar Harbor Formation

A cliff walk east of the town dock in Bar Harbor leads to nearly continuous outcrops of sandstone and siltstone of the Bar Harbor formation. The rocks vary in color from brown through red and pink to white. Look for the remarkably uniform beds that grade from coarse at the base to fine at the top, evidence that they were deposited from turbidity currents. They dip down to the north, seaward. And be sure to watch for the large boulder of granite balanced on an outcrop of Bar Harbor formation. The glacier of the last ice age brought it here from the Cadillac pluton.

You can see other exposures of the Bar Harbor formation at the Ovens on the northeast part of the island, on the south side of the Porcupine Islands off Bar Harbor, and near Northeast and Southwest Harbors on the south side of the island. The Bar Harbor formation was probably deposited during late Silurian time. It contains volcanic ash, which probably erupted from the chain of volcanoes that formed parallel to the trench that was then swallowing the floor of the Iapetus Ocean. The Avalon terrane finally welded to North America when the last of that ocean disappeared, during Devonian time.

Cranberry Island Series

Rocks of the Cranberry Island series lie on top of those of the Bar Harbor formation and look very much like them. They were deposited during late Silurian or early Devonian time and resemble rocks on Penobscot Bay named the Castine formation.

The volcanic rocks of the Cranberry Island series are restricted to the southern portion of Mount Desert Island, particularly on the Cranberry Islands and the broad peninsula south of Southwest Harbor. The rocks are rhyolites and tuffs laid down from volcanic ash flows and interbedded with slates and siltstones. Many of the volcanic rocks formed during explosive eruptions. Good exposures exist near the picnic area at the Seawall Campsite. Some of the rocks exposed at Otter Cliffs belong to this volcanic rock series.

Cadillac Granite and Cadillac Mountain

Cadillac Mountain, at 1,530 feet, is the highest peak on Mount Desert Island, probably the highest on the west coast of the Atlantic north of Rio de Janeiro. The morning sun shines on the summit of Cadillac Mountain earlier than on any other place in the United States during the fall and spring. Pink granite of the Cadillac pluton underlies nearly half of Mount Desert Island, including Cadillac Mountain. The rock owes its pink color to its orthoclase feldspar. Its coarse texture makes the grains of feldspar, glassy gray quartz, and black hornblende easy to distinguish. A plaque about 100 yards toward Bar Harbor from the summit parking area shows an enlarged view of these mineral grains. Nearly the whole summit of Cadillac Mountain consists of outcrops of the Cadillac pluton.

The granite has prominent joints, or fractures, both vertical and parallel to the surface. Those parallel to the surface are called sheeting joints; they form as the rock expands as erosion removes the overburden. Sheeting joints help determine the shape of the mountain. Disintegration of the rock along them creates small openings that make them very obvious.

You can see a 360-degree panorama of Mount Desert Island and its surroundings from the summit of Cadillac Mountain. To the east and northeast is Frenchman's Bay, which leads to Bar Harbor, named for a sand spit that connects to Bar Island at low tide. A breakwater connected to Bald Porcupine Island protects the southeast side of the harbor.

The Porcupine Islands are so named because the pointy spruce trees on the glacially rounded summits of several suggest porcupines. All are eroded from a sill of diorite of Devonian age that intruded the Bar

Weathered and eroded joints in Cadillac granite on the auto road near the summit of Cadillac Mountain, Acadia National Park.

Harbor series of volcanic sediments. The sill, which dips gently to the north, fractures into columns. Look for them on the steep south side of each island, where they make cliffs, which the plucking glacial ice may have steepened.

Look south to see the Cranberry Islands and Northeast Harbor. The lowland of the southern Maine slate belt lies to the north on the far side of the Norumbega fault. The highest point in that direction is Peaked Mountain in the Lucerne pluton, unless Mt. Katahdin is visible. Trees on the summit block the view to the west. The highest turnout below the summit gives an uninterrupted view west of Blue Hill, with the Camden Hills beyond. Far below the road is the deep glaciated valley occupied by Eagle Lake.

Shatter Zone

An interesting feature called the shatter zone surrounds the main portion of the Cadillac granite pluton. It is a breccia, fragments of the Bar Harbor, Ellsworth, and Cranberry Island groups of rocks, all embedded in the Cadillac granite, which intruded them. The intruding magma shattered these rocks, then filled the spaces between the blocks with pale gray granite.

Such forceful shattering of the surrounding rock during intrusion of a pluton is not common in Maine, where most granites have a sharp and unbroken contact with their enclosing country rock. No one has offered an entirely convincing explanation for the origin of the shatter zone. Perhaps part of the magma erupted, leaving the ground above without support. It collapsed into the magma chamber, creating the shatter zone at depth and opening a caldera basin at the surface.

A number of good exposures of the shatter zone exist along the Park Loop Road. Watch for them at the first turnout south of the park headquarters, on the east side of Sand Beach, and at Otter Cliffs.

Somesville Granite

The Somesville granite was named after Somesville, the oldest permanent settlement on Mount Desert Island, dating to 1761. This pink to gray granite appears toward the west side of the island, from Somes Sound almost to the western shore. It must be younger than the Cadillac granite because it cuts through the shatter zone near Town Hill Corners, north of Somesville. The Somesville granite is finer grained than Cadillac granite, and it contains two kinds of feldspar: pink orthoclase and gray plagioclase. The black speckles are biotite mica, rather than hornblende. These distinctions are hard to see without a good magnifier.

Blocks of Cranberry Island volcanics and Ellsworth schist in the shatter zone of the Cadillac granite, Sand Beach, Acadia National Park.

The coast of Maine has long been noted for its granite quarries. They were close to water, and thus the granite could be shipped by sea. On the west side of Somes Sound, near the head, a number of quarries operated from the late 1800s well into the 1900s. The village of Hall Quarry was at the center of the quarrying, apparently named for one of the larger granite operations. It supplied Somesville granite for many buildings along the East Coast.

Schoodic Peninsula and Isle au Haut

The Schoodic Peninsula and Isle au Haut are part of Acadia National Park but not of Mount Desert Island. Both areas are geologically similar to the rest of the park.

The Schoodic Peninsula is composed mostly of Gouldsboro granite, similar to the fine-grained pink rock near Southwest Harbor. The several dikes of basalt and rhyolite that cut the granite seem related to the mixed magmas of the Bays of Maine igneous complex. In any case, they intruded fractures in the older rocks during Devonian time. The best exposures of these dikes are near the parking lot, especially to the west, on Schoodic Point, which is actually the south end of Big Moose Island. Dikes of black basalt about 3 to 10 feet across cut through the pink granite. They trend generally north. Several dikes intersect; see if you can tell which are the younger.

One dike has an especially conspicuous border of finely crystalline rock that grades toward the center into a much coarser version of basalt called diabase. The finely crystalline border cooled against the molten, but much cooler, granite magma. The coarser basalt within the dike cooled slowly enough to permit larger crystals to grow. One wide dike of pale rhyolite is banded, and contains a scattering of oversized crystals of glassy quartz and milky rhyolite. Like the other dikes, it was probably emplaced during Devonian time, some 380 or so million years ago.

Rocks at the northern end of Isle au Haut are metamorphosed mudstones similar to the Ellsworth schist. The colorful volcanic rocks of the Cranberry Island series lie on them. Granite and gabbro intrusions of Devonian age intrude the Cranberry Island series, as well as the older rocks beneath.

Contact between pale Gouldsboro granite and dark basalt dike, Schoodic Point, Acadia National Park. The basalt under the coin cooled quickly to fine-grained rock, while the interior formed larger crystals upon cooling more slowly.

Somes Fjord and Other Glacial Valleys

Somes Sound is said to be the only fjord on the East Coast of the United States. It is a long and more or less straight valley that nearly divides Mount Desert Island. Glacial ice gouged it out, and the rising sea levels of postglacial time flooded it.

A number of other long and narrow glaciated valleys hold lakes, including Eagle Lake, Jordan Pond, Bubble Pond, Echo Lake, Long Pond, and Seal Cove Pond. These lakes are more closely oriented with the directions of sets of fractures in the bedrock on Cadillac Mountain than with directions of ice flow. I believe that rivers eroded their valleys along these prominent sets of fractures before the ice ages began. Then the great glaciers straightened the river valleys and gouged them into broad troughs.

Sand Beach

Sand Beach is one of several unusual beaches on Mount Desert Island. About 70 percent of the sand grains are shell fragments; the remainder are quartz and feldspar. The shell fragments are mostly from greenish sea urchins and pink and white mussel shells, all washed in from the productive waters offshore. The quartz and feldspar came from nearby granite bedrock and from glacial deposits.

It is unusual to find so much shell fragment sand in a beach on such a cold coast. The shells are composed of calcium carbonate, which is rather easily soluble in cold water. Carbonate beaches can exist near Mount Desert Island because the nearby waters are so extremely productive that shell fragment sand forms faster than the fragments dissolve. You would have to travel south to Georgia to find another beach with as much calcium carbonate as is found on Sand Beach. Geologists generally assume that ancient limestones, formed through cementation of carbonate beaches, are evidence of a warm climate. The carbonate content of Sand Beach and the bitter cold water there suggest that this assumption is not entirely correct.

A good wave-eroded outcrop of the shatter zone borders the Cadillac granite on the east end of Sand Beach. Also at the east end of the beach, a freshwater stream flows across the sand in wet weather. In dry weather, when stream flow is low, the water percolates through the sand. At low tide, the groundwater issues from the beach in a series of springs that make the sand look like it is boiling.

Thunder Hole

Thunder Hole is a narrow chasm that the waves eroded along a zone of fracturing in the bedrock. When waves break in this narrow

Dark basalt dike of Devonian age intrudes light Gouldsboro granite, also of Devonian age, on Scoodic Point, Acadia National Park. This dike trends away from the coast, while Mesozoic basalt dikes to the southwest run parallel to the coast.

opening, they trap air beneath them, which then bursts out, explosively. The optimum conditions for good explosions are a rising tide with a strong ocean storm offshore. At such times, however, footing along the walkway next to Thunder Hole becomes hazardous, as water shoots all the way back to the Park Loop Road, with thunderous noise and vibrations underfoot. Most of the time the action is so quiet that visitors wonder why it is an advertised stop on the Loop Road. The difference in levels of activity illustrates why geologists believe that virtually all wave erosion of coasts happens during occasional heavy storms.

Otter Cliffs and Otter Point

Rocks exposed in the steep cliffs at Otter Cliffs and Otter Point are of vertical beds of the Cranberry Island volcanic rocks. They are found in bright shades of pink, purple, and green. Individual layers do not continue far because they are much broken along faults. The rocks broke along a set of nearly vertical fractures to form the steep cliffs.

Seawall Beach

The gravelly Seawall Beach is more typical of this section of the Maine coast than are the sandy beaches of Cranberry Island and Sand

Beach. The pebbles on Seawall Beach are granite, pieces of the shatter zone, and various exotic rocks that the glaciers brought in from some distant source and then left in the glacial deposits. The pebbles are smooth and nicely rounded, silent testimony to the noisily vigorous wave and tidal conditions on these shores. Heavy waves breaking high on the beach built the high beach ridge during strong ocean storms. During all but the most severe storms, that beach ridge acts as a natural seawall, hence the name.

U.S. 1
Ellsworth—Calais

121 miles

See map on page 76

Ellsworth Moraine

A few miles east of its junction with the road to Bar Harbor, U.S. 1 climbs the steep slope of a glacial moraine deposited in seawater. At the top, you can see a cemetery, pine and blueberry barrens, and several pits in the gravelly sand. All are common features of sandy marine moraines. The road loops along the crest of the moraine for a few miles, mimicking the lobate front of the glacier that deposited it.

Views from Sullivan and Hancock

Many people who know something about scenery consider two views of Mount Desert Island from U.S. 1 as beautiful as any in the world. One is in Hancock, at an advertised roadside turnout, the other in Sullivan, just west of the junction with Maine 200. Both places offer spectacular views to the south across Frenchman's Bay to the mountains of Mount Desert Island.

Frenchman's Bay was eroded mainly in the Ellsworth schist. Volcanic rocks of the Bar Harbor series thinly cover it in some areas.

Pineo Ridge

Near Cherryfield is a large complex of glacial deltas and moraines that was deposited near the inland limit of marine submergence at the end of the last ice age. Large areas of outwash seaward of the moraine were reworked into beaches as the continental crust rose and the sea retreated. The glacial deposits also include some of the longest eskers

in the world, relics of meltwater streams that drained from the base of the glacier. Two of the eskers have been traced, with some breaks, northwest to the vicinity of Katahdin. As is usual with large volumes of sand in Maine, it probably came from nearby masses of granite, in this case from the large plutons northwest of Pineo Ridge.

The glacial landforms are easy to see because blueberry fields cover them, rather than trees. Washington County is the blueberry capital of Maine and the Northeast, specializing in "wild" blueberries. These low bush plants resist fire; the fields are routinely burned in the fall and produce again two years later. The fires inhibit the growth of plants that complete with the blueberry plant. Harvesters gather the berries with a rake—a special device that vaguely resembles a dust pan, only deeper, and with stiff wire teeth on the bottom. The rakers move on their knees along long rows marked by string, picking up the berries and a few leaves with a scooping motion of the rake. The fifty to one hundred dollars that a good raker may make in a day is earned by backbreaking labor.

Denbo and the Great Heath

The Great Heath north of Pineo Ridge contains a large accumulation of peat. Deposits of glacial sand and mud on the moraine impounded surface water, initially forming a pond. Decaying vegetation gradually filled the pond, converting it into the largest raised bog or heath in the state.

An electrical generating plant recently went into production, burning peat mined from a nearby bog in Debois called the Denbo Heath. The original plan was to compact the peat and extrude it into pellets that would fire the plant, but it proved difficult to dry the pellets. The plant now runs on a mixture of fluffed peat and wood chips.

Tunk Lake Pluton

Maine 182 is a popular shortcut between Franklin Road and Cherryfield. It runs somewhat north of U.S. 1. Large boulders of granite eroded from the Tunk Lake pluton occur along sections of the road. Some are in moraines, others simply rolled down from cliffs above the road.

Catherine Mountain, between Fox Pond and Tunk Lake, was reported in 1920 to have the world's largest deposit of the mineral molybdenite. Molybdenum is an essential ingredient in alloy steels. If this deposit were as good as advertised in 1920, mining would have started long ago.

Internal structure of DeGeer moraine at Addison. Overriding ice folded outwash beds into a syncline and left the juicy till that covers them.

Addison Moraine

A pit about 50 feet deep exposes the anatomy of a DeGeer moraine in detail so marvelous that it borders on indecency. You can see ridges of debris that the glacier deposited while it was grounded on the seafloor in perhaps 200 feet of water. Meltwater flowing on or within the ice carried gravel and sand to the edge of the glacier, where it was grounded. Glacial mud shed from the floating ice and from icebergs slowly settled on the submerged moraine.

Ice rode over the Addison moraine, folding some of the gravel beds into a syncline that lies on its side like a tired jelly roll. Muddy glacial till scraped off the base of the moving glacier covers the whole contorted mess.

Between Columbia Falls and Machias, U.S. 1 follows the crest of a long moraine with an origin similar to that of the Addison moraine. The broad sweeping curves in the road follow the rounded lobes of the moraine—the embayed margin of the glacier.

Machias

A trading post was established here in 1633; loggers founded the town in 1763, a year after wildfires from New Hampshire swept the coastal forests from Portland southward. The loggers were looking for masts for sailing ships and found the long pine along the Machias River. In 1775, less than a month after the Battle of Lexington, the first naval engagement of the Revolution took place off Machias, in which the locals captured a British convoy seeking lumber and sank a British schooner, the *Margaretta*. Artifacts from this engagement are on display at Burnham Tavern in Machias, said to be the oldest building in Maine east of the Penobscot River and the place where the naval skirmish was planned.

Around 1700, the pirate Samuel Bellamy established a fort in Machias in which to store his booty. He soon afterward attacked a French naval ship by mistake; it chased him south to near New Bedford, in Massachusetts. Bellamy then captured a whaling ship there, whose captain pretended to join him in his pirating. The whaling captain intentionally ran Bellamy's ship, the *Whidaw*, aground near Cape Cod, with the loss of all hands and the pirate booty. In the mid-1980s, divers recovered some of the gold and other artifacts.

Near East Machias, U.S. 1 passes high outcrops of Silurian volcanic rocks, part of the coastal volcanic belt. The rocks are mostly explosively erupted fragmental rhyolite, along with a few basalt flows. All were folded when the Avalon terrane collided with North America during the Acadian mountain-building event.

Fragmental volcanic rocks of Silurian age in East Machias. Similar rocks on Mount Desert Island and to the east to Quoddy Head are part of the coastal volcanic belt.

Jasper Beach

A small and steep beach at Bucks Harbor, south of Machias, contains nicely rounded and polished pebbles of volcanic rock, striped in shades of brown, red, and black. They came from outcrops of the Eastport formation, which is Devonian in age, at the south end of the beach. The few pebbles of granite were eroded from the bank of glacial till on the south end of the beach. The beach face is steep mainly because most of the water that waves cast onto it soaks into the gravel, leaving little to run back down the beach to move the pebbles.

At low tide the lower portion of the beach often assumes a scalloped appearance, with the sharp points toward the ocean. Beach cusps are said to form when waves that run parallel to the beach, called edge waves, interfere with storm waves approaching the shore straight on. The spacing of the cusps is related in part to the distance between the crests of the approaching waves.

Farther south of Bucks Harbor is the small village of Starboard. On the pebbly beaches at the end of the road are small outcrops of purple shelly limestone that belong to the Silurian Hersey formation.

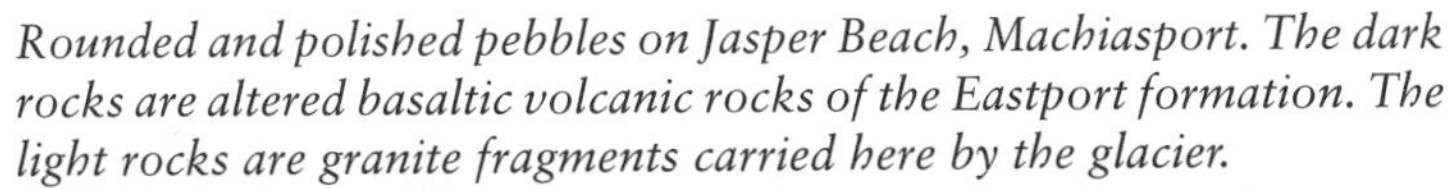

Rounded and polished pebbles on Jasper Beach, Machiasport. The dark rocks are altered basaltic volcanic rocks of the Eastport formation. The light rocks are granite fragments carried here by the glacier.

Beach cusps exposed at low tide at Jasper Beach, Machiasport.

Cutler

Maine 191 heads south at East Machias toward Cutler and some of the least developed shore frontage in Maine. You can see a number of DeGeer moraines along the way, and a group of very-low-frequency radio antennas. The Navy uses this system of antennas, with a length of 20 miles, to communicate with submarines that are close to the surface. An electromagnetic field around the antennas transmits a signal through the earth as well as the ocean. This very-low-frequency signal can be received as far away as South Africa and Alaska. Geologists have made use of this, and similar signal sources in Maryland and Washington, to detect faults and plumes of contaminated groundwater, among other things.

When the antennas were under construction in 1959, an excavation in the glacial marine sediments uncovered two walrus skulls. They are about 13,000 years old, the time when seawater flooded coastal Maine as the great glacier melted. The skulls are now on display, along with a mammoth tusk found the same summer, in the State Museum in Augusta.

The Cutler Radio Base was built after Project Sanguine failed. Someone, I hope not a geologist, thought that bedrock could be used as a radio antenna. Holes were drilled about 30 miles apart in both Cape Cod and Wisconsin and an attempt was made to generate a radio signal using the earth's crust as an antenna. The Navy had been told that because the granite at the two sites was uniform in all directions, it would act like a radio beacon. Any geologist would know that no rocks are uniform for any great distance, certainly not granite, and certainly not for miles.

Quoddy Head

West Quoddy Head is the easternmost point in the United States. The bedrock is an intrusion of dark gabbro that arrived as molten magma during Silurian time. It is called the Cutler pluton. It may have been the source of some of the basalt in this area.

The tidal currents between Quoddy Head and Campobello Island to the east are strong. They produce what is said to be the largest whirlpool in the Western Hemisphere near the northwest end of the island. Indians traveling in canoes along the coast preferred to carry across the narrow neck that connects Quoddy Head to the mainland. They dragged their canoes across Carrying Place Bog, through which the Quoddy Head road now cuts.

Quoddy Head, Lubec, the easternmost point in the United States. The narrow neck that connects Quoddy to the mainland is covered by the Carrying Place Bog. —Sewall Company photo

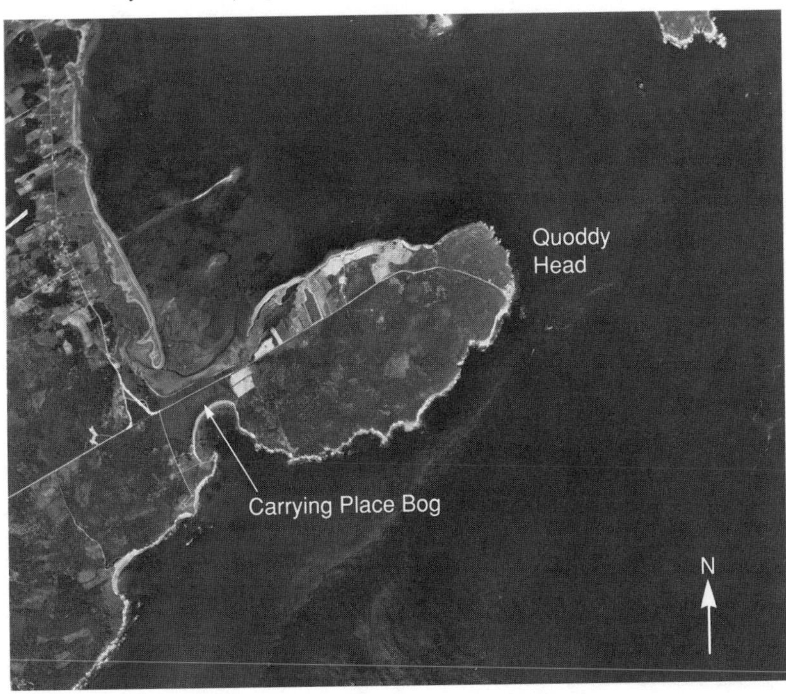

Carrying Place Bog is a raised bog, or heath, that supports a number of plants native to the arctic. The cold climate allows them to survive. The pitcher plant and the sundew also thrive; they are insect-eating plants that grow in many bogs in northern Maine. The pitcher plant has a long vertical tube that catches water to drown bugs that blunder in. The leaves of the sundew are hairy pads that open flat and secrete an attractive sugary fluid. When a bug lands to investigate, the pad closes around it and digests it.

Silurian and Devonian Volcanic Rocks

Many of the roadside outcrops between Machias and Calais are volcanic rocks assigned to a number of formations of Silurian age, as well as a few of Devonian age. They include black basalts and pink, maroon, and green rhyolite. All appear to have erupted into the ocean during a plate collision, with the water becoming shallower as more rock accumulated. In the northern woods of Maine, where I have worked and where exposures of rock are poor and far apart, I would likely have mapped all of these formations as a single but variable unit. I would be called a lumper. Here in the blueberry barrens and coastal cliffs where rocks are better exposed, people tend to be splitters. They separated similar looking rocks into five or six formations.

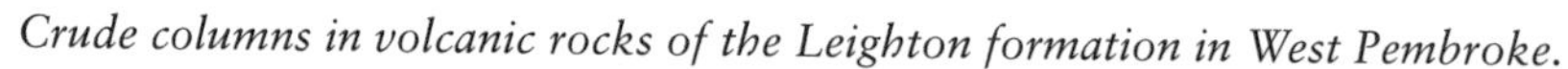

Crude columns in volcanic rocks of the Leighton formation in West Pembroke.

Basalt flows of the Eastport formation in Perry.

Watch at West Pembroke, at the intersection of U.S. 1 and Maine 214, for crude columns in volcanic rocks of the Leighton formation of Silurian age. Devonian basalt flows of the Eastport formation are exposed at Perry, on U.S. 1.

Perry Formation

The late Devonian Perry formation contains conglomerate and sandstone that were deposited on river floodplains; the sediments came from the Acadian mountains. It is one of the youngest sedimentary rocks widely exposed in Maine. Reddish pebbles of basalt from the Eastport formation and of the Red Beach granite dominate the conglomerates in the Perry formation, clearly establishing that it is younger than those rocks.

At Lamb Cove in Robbinston, at low tide, the Perry formation is exposed above a Devonian erosion surface developed on the Eastport formation. This buried erosion surface is called an unconformity; it represents a discontinuity in the geologic record, in this case one of some 30 million years.

The eminent Canadian geologist Sir John Dawson found fossils of primitive land plants in the Perry formation in the 1860s. They firmly

Sandstones and conglomerates of the Perry formation at Lamb Cove, Robbinston. The Eastport basalt is exposed on the right, on the beach.

establish its age as late Devonian, postdating all of the Acadian mountain-building event. Early Devonian plant fossils are found in the Trout Valley formation in Baxter State Park; middle Devonian plant fossils occur in the Mapleton formation near Presque Isle in Aroostook County.

Red Beach Granite

The Red Beach granite is exposed at Robbinston and near Red Beach, in the southern part of Calais. It is one of the numerous plutons that invaded the rocks of this region during Devonian time. It outcrops in an area of about 12 square miles near U.S. 1. It is a beautiful red rock, used to face two wings of the American Museum of Natural History in New York City.

The Red Beach granite west of the highway is gray. It becomes red or purple only near its contact with the Perry formation. The Red Beach granite intruded the early Devonian Eastport formation, an assemblage of volcanic rocks, mostly basalt and andesite.

By late Devonian time, about 30 million years after the intrusion of the Red Beach granite, erosion had exposed both the Eastport formation and the Red Beach granite, and soil had developed on their upper surfaces. Beneath the soil, the Eastport formation is black and the Red Beach granite is gray. The granite and the flows in the Eastport formation owe their color to the mineral hematite, red iron oxide, from the late Devonian soil. Such red soils form today only in wet tropical or subtropical climates. Evidently, the climate of Maine was like that during late Devonian time.

St. Croix Island and the St. Croix River

St. Croix Island is offshore from the landing and picnic area in Red Beach. The French explorer Samuel de Champlain named it in 1604, and the river that separates Maine from New Brunswick was named after it.

Rocks on St. Croix Island are the Red Beach granite, which has two prominent sets of parallel fractures that intersect at about right angles. Wave erosion along the fractures shaped the island like a cross, which inspired Champlain's name.

The fleeting French settlement on St. Croix Island staged the first Christmas celebration in New England in 1604. The treaty that ended the American Revolution in 1783 vaguely defined the border between Maine, then part of Massachusetts, and Canada as "that Angle which is formed by a line drawn due North from the source of the St. Croix River to the Highlands."

These highlands are the drainage divide between streams that flow north to the St. Lawrence River and those that flow south to the St. John River. Maps and geographic knowledge of interior Maine were primitive in 1783, so several years passed before the British and Americans could agree upon which was the St. Croix River. Champlain's map and some artifacts found on the island finally settled the matter.

Oak Bay Fault

Oak Bay is a small community east of St. Stephen, New Brunswick. Extending southeast from Oak Bay for about 40 miles is a remarkably straight embayment eroded along the Oak Bay fault. Occasional small and moderate-size earthquakes occur along this fault. Since the eastern part of Maine is now sinking at a rate of about 3 feet per century, it is easy to suppose that some of this movement might take place along the Oak Bay fault.

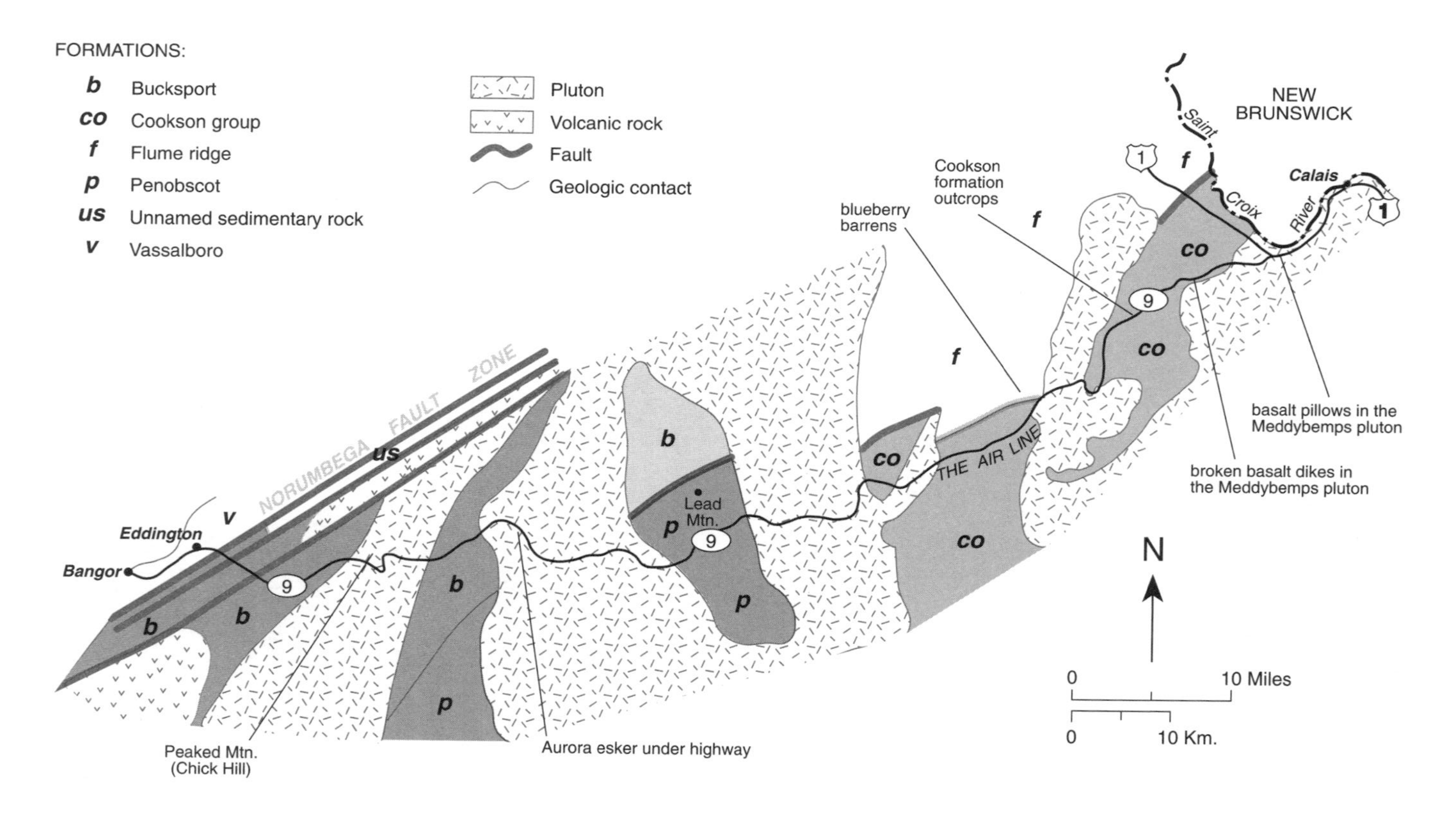

Bedrock geology along the Air Line between Bangor and Calais.

Maine 9
The Air Line
Bangor—Calais
98 miles

Maine 9 runs more or less from east to west between Bangor and Calais. Construction began in 1839 to speed troops to Calais during the Aroostook War, but the fracas ended before the road was finished. It was finally completed in 1857 as the route of a mail and passenger stagecoach service known as the Air Line because it was a direct route between Bangor and Calais, a day shorter than the Shore Line that followed the coastal route.

Afraid that the Air Line would attract its passengers, the Shore Line ran an ad in a New York paper showing wolves chasing the Air Line stage, hoping it would scare away business. Instead, many intrepid travelers chose the Air Line, hoping to get a chance to shoot wolves. The owner of the Air Line bought the woodcut of the wolves, used it in his own ads, and the line became known as the Wolf Route. Steamboats put the Air Line out of business in 1887. In the 1920s, mail planes used the Air Line to guide them between Bangor and Calais, which led many people to believe that is why it received its name.

Wolves attacking the Air Line stage.

The two towns at the eastern end of the Air Line, Alexander and Baring, are the first and last names of Lord Ashburton. He and Daniel Webster settled the dispute over the border between Maine and Canada in 1842.

Plutons of igneous rock that rose as molten magma during Devonian time underlie about half the Air Line. They make vast stretches of granite and gabbro. Small areas of the early Paleozoic sedimentary rocks that they invaded separate them.

Norumbega Fault Zone and Post-Acadian Sedimentary Rocks

Just east of Eddington, the Air Line crosses the Norumbega fault zone, which separates the Avalon terrane from the Central Maine slate belt. The Norumbega fault zone contains at least three parallel faults, between two of which is a thin slice of conglomerate and sandstone that were deposited during late Devonian or early Carboniferous time, about 360 million years ago. Streams flowing along the fault zone may have deposited them. In any case, these are the youngest sedimentary rocks in Maine.

Horizontal movement along the Norumbega fault zone broke these sedimentary rocks, so the fault has been active since they were laid down. That movement may have happened as Africa collided with North America during Upper Carboniferous time, in the Alleghanian mountain-building event. If so, it happened sometime around 320 million years ago.

New Brunswick has large areas of sedimentary rocks about the same age as these. It seems likely that these slivers of sandstone and conglomerate are the remains of a much wider blanket of rocks now mostly lost to erosion. Other small slices of similar rock exist within the Norumbega fault zone near the New Brunswick border. Also, a block of highly metamorphosed rocks occurs near the border, apparently moved up the fault from greater depths.

Lucerne Granite

The western portion of the Air Line crosses a broad area of the Lucerne granite, noteworthy for its large crystals of pinkish feldspar. Each one nestles within a halo of black crystals of hornblende and biotite mica.

Quarries in the Lucerne granite near the town of Lucerne produced some granite doorsteps that had an interesting history. Several were installed in front of a building in Bangor that later became a brothel. It was closed about 1971, when the log drives ceased on the Penobscot River. Loggers rescued an iron support beam and a granite doorstep during demolition and incorporated them into the loggers' memorial on Chesuncook Lake. Tools of the loggers' trade were welded to the metal post, with the door step placed in front. It remains a popular tourist stop in the Maine woods, although few know the origin of the support post and doorstep, or its association with the Lucerne pluton.

Peaked Mountain is the highest elevation in the Lucerne granite near the road. The short trail to the summit provides a commanding view of a broad expanse of eastern Maine. On clear days, the mountains on Mount Desert Island rise far to the south and Mt. Katahdin is visible in the northwest.

Aurora Esker

The major esker systems of eastern Maine extend northwest as far as Katahdin. Many elementary geology laboratory manuals illustrate eskers with a map of the region around Enfield, about 25 miles northwest of Aurora. The Air Line crosses that esker system; several miles of the road follow its crest, the Whaleback. This esker is unusual for Maine because it is a straight ridge for about 3.5 miles, rather than

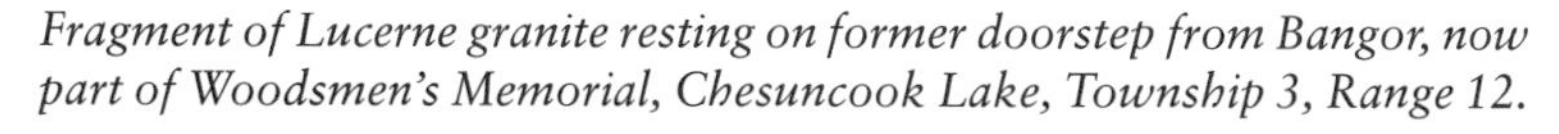

Fragment of Lucerne granite resting on former doorstep from Bangor, now part of Woodsmen's Memorial, Chesuncook Lake, Township 3, Range 12.

winding back and forth like a snake. The straight segment was likely deposited in a fracture open to the air, rather than in a tunnel.

The same esker system about 10 miles to the northwest, in Township 32, is known as the Horseback, the other common Maine name for these ridges. This esker records the location of one of the meltwater streams that fed sand and gravel to the Pineo Ridge marine delta, about 15 miles to the south.

The word *esker* comes from a Celtic term for a pathway or trail. The swamps and ponds that border the Aurora Whaleback make it the obvious route for this part of the Air Line. Other local names for eskers include Indian Trail and Indian Railroad. Evidently, the Celtic people of Europe and the Indians came to the same conclusion about the usefulness of eskers.

Lead Mountain

About halfway between Bangor and Calais on the Air Line, Maine 193 leads south to Cherryfield. Just east of this intersection, on the west side of the Narraguagus River near a Maine Forest Service camp, a trail leads north about 2.5 miles to the summit of Lead Mountain, or Humpback as the locals call it. The bedrock is the Penobscot formation, of early Paleozoic age. This rusty rock contains small blebs of the mineral pyrrhotite, which may have been mistaken for lead.

Iron sulfide–rich rocks of the Cookson formation, Alexander.

Pocomoonshine Pluton

Pocomoonshine is the name of a small lake north of the Air Line; the word may derive from the Algonquin for "broken-off-smooth." The Pocomoonshine pluton is a mass of gabbro to diorite, dark rock that was emplaced as molten magma during Devonian time. The heat of the magma baked the surrounding muddy sediments of the Cookson and Flume Ridge formations, converting them into metamorphic rocks. They resist erosion and stand slightly higher in the hills around the less resistant rocks of the pluton, especially along its western margin. Much of the pluton has weathered to low ground occupied by swamps and Pocomoonshine and Crawford Lakes.

Meddybemps Pluton

Meddybemps is the name of a lake, presumably taken from the Algonquin word for "plenty of alewives." In the 1920s, a series of liberal political tracts issued in Maine were called the Meddybemps Papers; Bowdoin College has a double quartet of singers called the Meddybempsters.

The Meddybemps pluton is near the eastern end of the Air Line, just west of Calais. Like many others in the eastern coastal zone of Maine, the Meddybemps pluton is a mixture of igneous rock types, including both granite and basalt. They date from Devonian time. Granite magma forms through melting of rocks of the continental

Disrupted basalt dike in the Meddybemps pluton, Alexander.

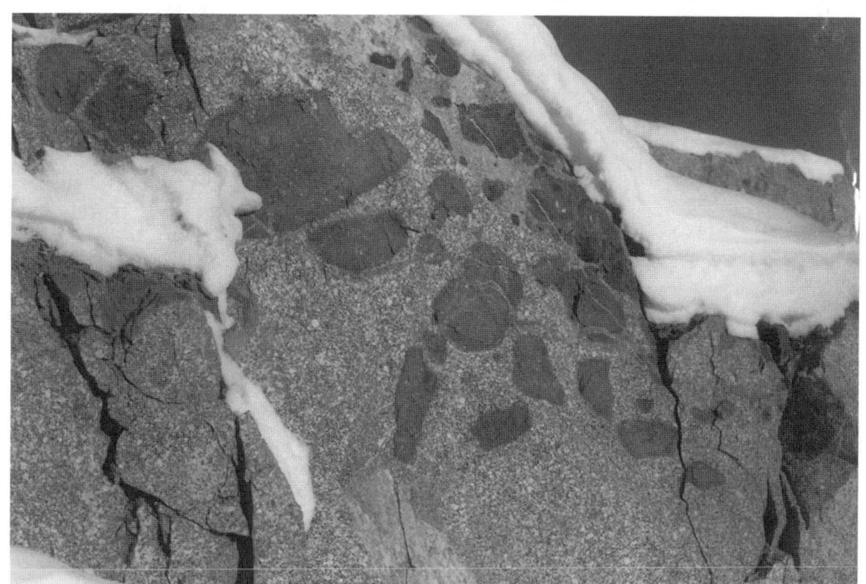

Basalt enclaves in the Baring granite on U.S. 1, near its intersection with the Air Line in Baring.

crust, and basalt magma melts in the mantle. How could these very different rocks that come from such widely separated sources find their way into the same igneous intrusion?

One theory now much in favor holds that liquid basalt intruded the granite while it was still molten. Basalt magma is nearly twice as hot as molten granite, so the basalt was quenched to pillow structures in the cooler liquid, as happens in seawater to make ordinary pillow basalts. In this case, layers of granite separate many of the pillows.

Watch for the excellent exposures of these unusual rocks along the trails in the Moosehorn National Wildlife Refuge. You can see similar rocks well exposed in outcrops of the similar Baring pluton in Baring on U.S. 1, just east of its junction with the Air Line.

Central Maine

Central Maine lies between the coast and the belt of mountains that trends northeast through the middle of the state. The Norumbega fault zone separates central Maine from the coastal region; the rocks on either side of the fault look different, and they have different geologic histories. The rocks in central Maine resemble those in the mountains more than those along the coast.

The availability of water power dictated the locations of most of the towns in central Maine. The larger cities, including Lewiston-Auburn, Augusta, Waterville, and Skowhegan, grew at waterfalls and rapids on the major rivers. Streams, many named Mill Brook or Mill Stream, were dammed to supply water for sawmills, gristmills, and other uses in many other smaller towns. Maine has twenty-three Mill Brooks, eleven Mill Streams, nine Mill Ponds, and six Mill Creeks. Maine also has sixty-five Mud Ponds, forty-six Long Ponds, and forty-six Meadow Brooks.

Central Maine Slate Belt

Although other types of sedimentary rocks exist, the characteristic rock of central Maine is slate. That is why the region is also called the Central Maine slate belt. It is a tract of slates, metamorphosed sandstones, limy shales, and in the southwestern part of the belt, schists and gneisses. The Central Maine slate belt is about 120 miles wide. For a number of reasons, the degree of metamorphism intensifies toward the southwest.

The rocks are more thoroughly recrystallized in the southwest partly because they baked in the heat of the large Sebago granite pluton and partly because they were more deeply buried. The high temperatures and pressures that prevail deep within the continental crust bake the rocks there, converting them into metamorphic rocks. The mountains that rose during the Acadian mountain-building event must have been higher in southwestern Maine than farther northeast. Mountains float like icebergs. As icebergs melt from above, ice rises from below. As mountains erode, rocks rise from below. So it is that the once deeply buried rocks of southwestern Maine are now at the surface. While it

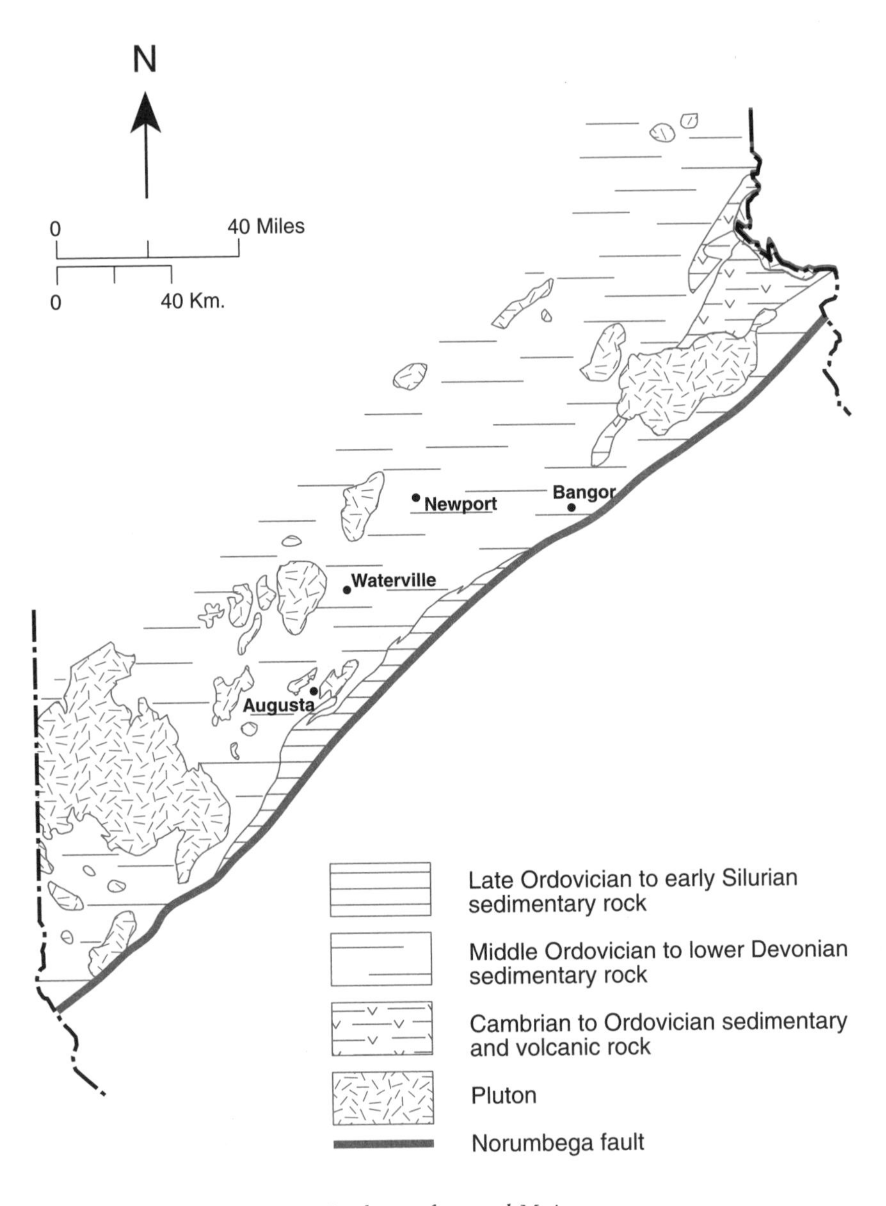

Geology of central Maine.

is obvious that these rocks were originally laid down as sediments, their metamorphism testifies to their sojourn in the depths.

Metamorphism produces new minerals characteristic of the temperatures at which the rocks recrystallized, and to some extent the pressures. The minerals in a metamorphic rock are the thermometers and barometers of its metamorphism. Green chlorite mica occurs in slates that have been buried at shallow depths of a mile or so. Black biotite mica and garnet crystallize in schists that recrystallized at higher temperatures, and generally at greater depth. Fine needles of sillimanite grow at even higher temperatures and pressures.

Thus it is possible to follow a particular rock layer from a slate in eastern Maine through the various metamorphic zones from green chlorite slates in the northeast to a sillimanite schist in the southwest. Some rocks on the southwestern side of the Sebago granite were subjected to even higher temperatures and became gneisses. Other studies have shown that the granites now exposed in southwestern Maine crystallized at greater depths than those to the northeast.

The rocks of the Central Maine slate belt are tightly folded into anticlines and synclines, arches and troughs. A broader look at the structure reveals that the small folds are superimposed on much larger super anticlines and super synclines, anticlinoria and synclinoria. An anticlinorium is a wide belt of complexly folded rocks in which the formations become generally older toward the middle. In a synclinorium, the formations become generally younger toward the middle. The Central Maine slate belt contains two synclinoria, the Kearsarge–Central Maine synclinorium and the Fredricton trough. The Miramichi anticlinorium separates them in northeastern Maine.

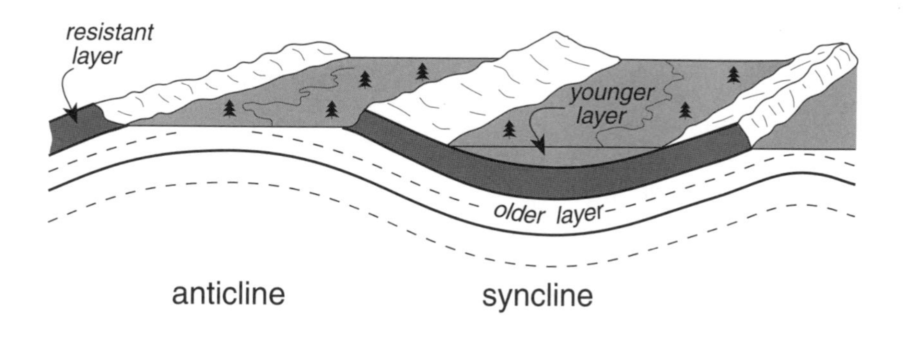

Anticline and syncline.

Events of Middle Ordovician and Late Silurian Time

After the Taconic mountain-building event of middle Ordovician time, the newly risen mountains shed large volumes of sediment east to the ocean. Those were coastal mountains. Meanwhile, in northern Maine, the newly attached Boundary Mountain and Gander terrane was breaking up. There, the coast faced a large sea that was separated from the ocean by a chain of islands. The islands were a displaced piece of crust that had been carried southeastward along faults. It eventually found its way into the Central Maine slate belt, where you now see it in the Miramichi anticlinorium, wedged between the Fredricton trough and the northern arm of the Kearsarge–Central Maine synclinorium.

Heavy, Muddy Waters

Rocks in the two great synclinoria began as muddy sediments deposited in deep ocean water. Most were dumped from turbidity currents, flows of water so muddy that it is heavy. Most turbidity currents probably start when a submarine landslide suddenly loads a large volume of sediment into the water. The dense mixture of mud and water flows rapidly downslope until it settles on the ocean floor. From middle Ordovician to middle Devonian time, turbidity currents blanketed the deep ocean floor off the mountainous coast with a muddy cover several miles thick.

Turbidity currents deposit a distinctive kind of mudstone in which each layer has the coarsest sediment at the base, grading to finer sediment at the top. Many are sand at the base, clay at the top. Each graded bed is the deposit of a single turbidity flow. Nearly all of the formations in the Central Maine slate belt consist mostly of mudstones laid down during the period from late Ordovician through early Devonian time, between about 450 and 370 million years ago.

Deep-Sea Fans Build Eastward

The southern edge of ancestral Maine, as it then was, faced the open Iapetus Ocean. Turbidity currents carried sediment into the basin, creating large fans that built out onto the deep ocean floor, becoming thinner to the east. The coarser sands and gravels accumulated in the channels that distributed sediment across the fan, while the finer silts and clays spread either across the fan surface or onto the deep floor of the Iapetus Ocean.

Meanwhile, the Avalon terrane was steadily approaching. The ocean floor that once separated it from North America lay beneath the sediments that you now see crumpled into the Fredricton trough and the southern arm of the Kearsarge–Central Maine synclinorium. That ocean floor is long gone, vanished into the earth's mantle. It slid out from beneath the sediments that once covered it, through an oceanic trench, and plunged into the depths. The sediments were crumpled into the folds you now see as they were jammed into the trench.

Fredricton Trough and the Miramichi Anticlinorium

The Fredricton trough, between the Miramichi anticlinorium and St. Croix belt, is a tract of tightly folded mudstones some 10 to 30 miles wide. The Norumbega fault zone defines its southern boundary against the much older rocks of North America.

Fossils in the rocks of the Miramichi anticlinorium and St. Croix belt more nearly resemble those in rocks of the same age in Europe than those west of the Norumbega fault zone. Geologists interpret the affinities of those fossils as evidence that the Iapetus Ocean separated them until it closed during the Acadian mountain-building event of Devonian time. The Miramichi anticlinorium contains volcanic and metamorphosed sedimentary rocks that were part of the Boundary Mountain and Gander terrane when it was on the other side of the Iapetus Ocean. It joined North America during the Taconic mountain-building event of Ordovician time. The St. Croix belt consists mainly of sandstones and of slates that weather to rusty colors. All were deposited along the margin of the Avalon terrane when it was still on the far side of the Iapetus Ocean.

Kearsarge–Central Maine Synclinorium

Mudstones continued to accumulate on the deep ocean floor after the Taconic mountain-building event. Submarine fans continued to build out from the old margin of North America until late Silurian time. Those sediments are in the Rangeley, Perry Mountain, Smalls Falls, Sangerville, Waterville, and Allsbury formations, among others. Meanwhile, on the other side of the Iapetus Ocean, an oceanic trench was steadily swallowing the floor of the Iapetus Ocean, which sank beneath the Avalon terrane and on into the earth's mantle. A chain of volcanoes erupted above the sinking ocean floor on the Avalon terrane. Some geologists believe that another trench on the North American side later swallowed some of the Iapetus Ocean floor during early Devonian time.

By the end of Silurian time, the floor of the Iapetus Ocean was mostly gone, and the continents that it once separated were colliding.

Sediments that had been spread across the ocean floor were shoved westward as the Avalon terrane approached the old margin of North America and rode over it. The North American continent sank beneath their immense weight in a depression that extended beyond the load of displaced sediments—if you press a soft rubber ball with your finger, the depression will extend beyond it. Sediments shed from the Avalon terrane washed into that marginal depression and filled it. You see them now in the Kearsarge–Central Maine synclinorium, most notably in the Carrabassett formation.

Sebago Pluton

The Sebago pluton dominates the geologic scene between Portland and the area north of the Androscoggin River, near Rumford. The Sebago granite was intruded during Carboniferous time as a horizontal sill, a layer of granite about a mile thick. Now it tilts down to the north at a gentle angle.

The rocks exposed south of the Sebago pluton originally lay underneath it. They were metamorphosed in its heat, in the normally higher temperatures at great depth, and under the pressure of the overlying rocks that have since eroded. The rocks exposed north of the Sebago pluton originally lay above it. They were metamorphosed at much shallower depth and under much less pressure. In general, rocks south of the Sebago pluton show greater metamorphic effects of pressure, while those to the north were influenced more by the rising heat than by burial at depth.

Oxford County Pegmatite Belt

In the last stages of its crystallization, the granite magma of the Sebago pluton generated pegmatite magma that invaded the surrounding rocks. Many of those pegmatites exist around the northern end of the Sebago pluton, some in the granite, some in the surrounding metamorphic rocks. These are fancy pegmatites, the kind that contain rare and precious minerals, as well as the usual giant crystals of quartz, feldspar, and mica. Dozens of minerals occur only in pegmatites, some of those only in the pegmatites of Oxford County.

Mining began in the late 1800s, mainly for feldspar and mica, with gems as a by-product. A few jewelry stores supported the operation of pegmatite mines to produce tourmaline and aquamarine. Feldspar went into pottery and tile, first near Trenton, New Jersey, and later in Ohio. Large sheets of mica were mined for windows in woodstoves.

In the stove window trade, mica acquired the name isinglass, after its resemblance to a gelatin made from fish bladders that was made into windows for carriages and early cars: "It's got isinglass curtains that roll right down." Later uses in vacuum radio tubes and electrical appliances were based on the good electrical insulating properties of mica. You still see the heating coils of electric toasters wrapped around thin sheets of mica. Feldspar and mica production ended in Maine in the 1960s, but most of the pegmatite quarries remain open for mineral collecting.

Sebago Lake

Sebago Lake, the second largest in Maine and the third largest in New England, lies near the southeast margin of the Sebago pluton. Its maximum depth, 316 feet, about 40 feet below sea level, is the greatest of any in New England. Most of the other lakes within the pluton are longer by far in the northwest to southeast direction than they are wide, apparently parallel to bedrock fractures and the glacial flow direction. Sebago Lake is nearly as wide as it is long, with its narrow part at the southern end.

Like other granite bodies in Maine, the Sebago pluton weathered to a sandy soil for thousands of years before the great glaciers came. They deepened the river valleys already eroded in these weak rocks, forming lake basins. Thick glacial deposits that contain large volumes of sand are associated with all of these large granite plutons. Lakes with large sandy beaches are nearly all within or near large granite intrusions.

As the ice melted at the end of the last ice age, seawater flooded the coastal region of Maine to the southern margin of Sebago Lake, where a large marine delta was deposited. As the ice melted back northwest from Sebago Lake, meltwater deposited more glacial outwash, in great quantity. A lingering mass of ice must have lain where Sebago Lake now is; otherwise, the basin would have filled with outwash. Sebago Lake is in fact a large glacial kettle lake, even though most people think of kettles as ponds.

As seawater flooded part of the Sebago basin, migrating Atlantic salmon were somehow isolated from other Atlantic salmon, and they could not interbreed. These salmon in Sebago Lake and three other Maine lakes (Green, West Grand, and Sebec) that were near the limit of seawater flooding developed into a new variety, the landlocked salmon. Louis Agassiz, the man who first understood ice ages, identified and named the landlocked salmon *Salmo salar sebago*, Latin for "jumping salmon from Sebago." These fish are now raised in hatcheries and stocked in many large cold-water lakes. Atlantic salmon live most of their lives in the ocean, then return to the rivers to spawn. Landlocked salmon live most of their lives in deep lakes, then return to the tributary streams to spawn.

Maine 5
Saco—Bethel
93 miles

Maine 5 winds back and forth across the Saco River, never getting more than 5 miles from it between Saco and Fryeburg. The Saco River originates in the White Mountains of New Hampshire near North Conway and flows into Maine at Fryeburg. The river's spring floods are fed by melting snows, and summer flows are sustained by the release of groundwater stored in thick and widespread glacial sand and gravel deposits. The good flow and the sandy banks and channel for swimming make the Saco a popular canoeing river. The river was cursed by the Indian Chocorua, who jumped to his death from the mountain named for him rather than be captured and killed by settlers. As he leaped, the curse he made stated that one white person would die in the Saco each year.

Maine 5 begins in coastal Maine and quickly crosses the Norumbega fault into central Maine proper. The metamorphosed sedimentary rocks start with a belt about 10 miles wide of the Silurian to Ordovician Vassalboro formation, an interbedded sandy and calcareous mudstone. The rest consist of very similar rocks of the Ringemere formation, said to be Devonian and Silurian in age. The road crosses intrusive igneous rocks in Saco, Fryeburg, and in the stretch between North Waterford and Bethel.

Eleven small Mesozoic plutons exist along or near this route, the first of which is cut by the Norumbega fault and by Maine 5. Geologists call them stocks, and most are very small stocks indeed, less than 4 square miles in area. A stock is a pluton exposed over less than 40 square miles. Most of the stocks are Jurassic and Cretaceous in age, while a couple are apparently Triassic, but all obviously postdate all the known collisional tectonic events. The stocks also are younger than the breakup of the supercontinent Pangea and the initial formation of the Atlantic Ocean.

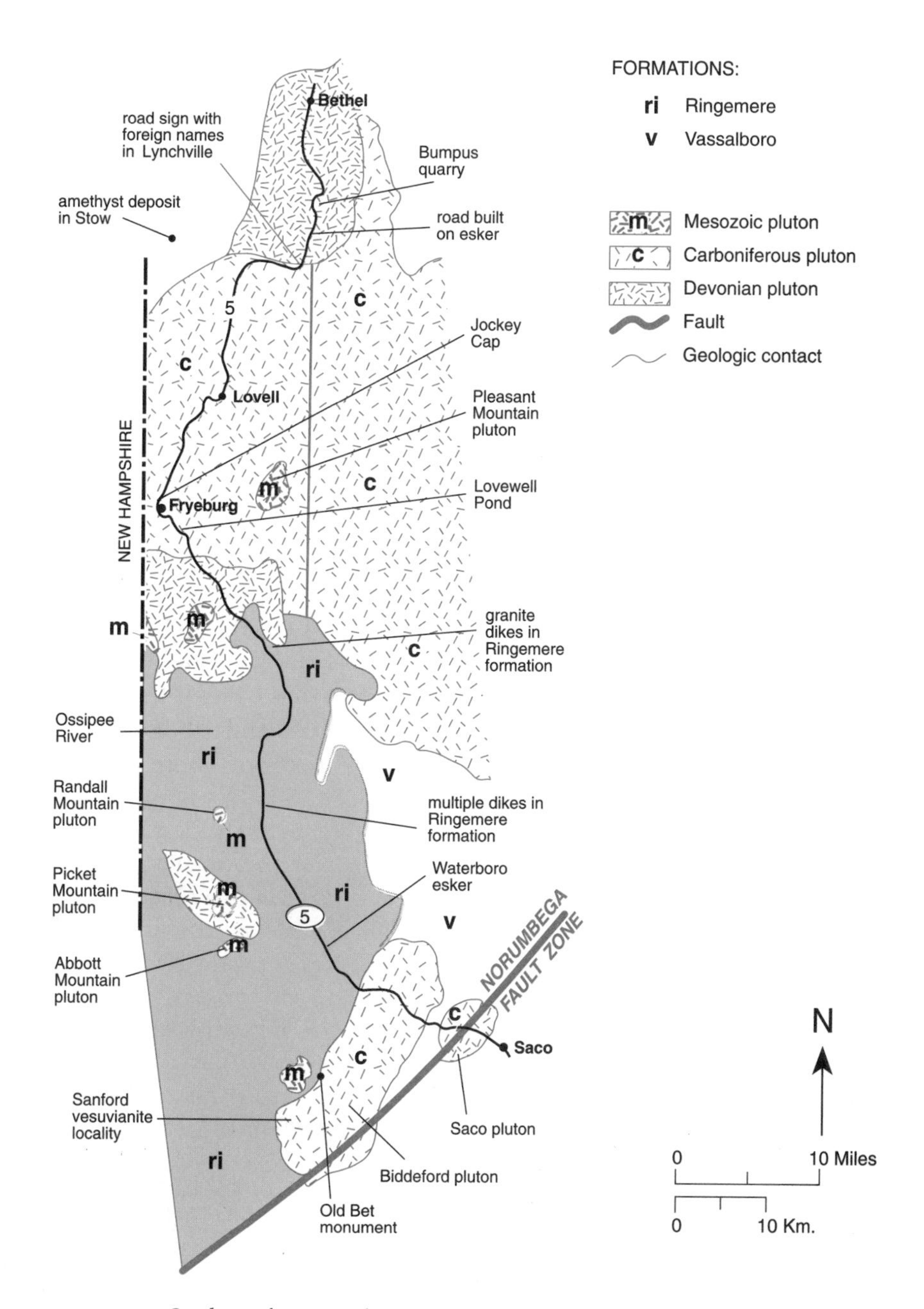

Geologic features along Maine 5 between Saco and Bethel.

Marine muds and extensive sand plains of the last ice age extend up the Saco River to near Cornish. Several eskers and outwash plains also occur above the limit of glacial marine flooding, giving the whole area a sandy appearance.

The Saco and Biddeford Plutons

Just beyond the crossing of I-95, the Carboniferous Saco pluton is cut and slightly offset by the Norumbega fault, indicating some young movements on that fracture. The rock is a gabbro. Only a couple of miles beyond is the Biddeford pluton, a Carboniferous granite, perhaps related to the nearby Sebago granite. The Norumbega fault breaks this granite, as well as the Saco gabbro.

The first crossing of many by Maine 5 over the Saco River occurs within the Saco pluton. The river here turns to the southwest for about 2 miles, exactly along the Norumbega fault. At least this is where geologists have located the fault on geologic maps. No outcrops are visible along this stretch of the river, but rivers always seek out the path of least resistance. Rocks along most faults are crushed and weakened, making them attractive routes for rivers.

High-Water Marks

South along Maine 35 about 2 miles from Maine 5 is an interesting opportunity to see the effect of the late glacial marine submergence

Saco gabbro on Maine 5 in Saco.

on some previously deposited glacial deposits. Look north of the road for a steep bank that waves cut into sand. Part of Maine 35 is on a beach. To reach this area, the water had to be about 220 feet higher than the present sea level.

Old Bet Monument

South of East Waterboro, on Maine 4 in Alfred, is a monument marking the shooting of a circus elephant, Old Bet. Her owner, Hackaliah Bailey, moved her from town to town at night and exhibited her during the day. A farmer objected to the money Bailey was taking from people during hard times and shot her as she passed through Alfred on July 24, 1816.

Sanford Vesuvianite Locality

About 3 miles south of the Old Bet monument is an interesting and famous mineral locality. In a wooded area off School Street, a narrow road leads a couple of hundred feet through a gap in a stone wall to a parking area. The mineral site is a short distance beyond, along a path through the woods. Two small openings have been blasted in the vesuvianite deposit.

The rock is a skarn, a zone of limestone metamorphosed by the heat of a nearby intrusion, in this case the limy sediments of the Ringemere formation intruded by the Biddeford pluton. The calcium carbonate in the limestone has combined with silica to form calcium silicate minerals, the components of a skarn. Vesuvianite, also called idocrase, is among the more rare calcium silicate minerals. The Sanford locality is listed in mineralogy textbooks as one of the prime occurrences in the world. Vesuvianite occurs in dark, lustrous brown stones, with striations along the long dimension of the crystals. This mineral was originally discovered in old lava flows on Mt. Vesuvius. Other minerals here are white calcite, green diopside, and reddish brown cinnamon garnet. Another interesting skarn mineral is scheelite, an ore of tungsten. At night an ultraviolet light causes the scheelite to glow a beautiful blue. That is how prospecting for this mineral is done.

East Waterboro Esker

Of the many eskers above the glacial marine limit in this area, one is crossed by Maine 5 in East Waterboro. This deposit was formed in one of several sub-ice channels that supplied sand and gravel to the Great Sanford outwash plain to the south. The Sanford sand plain is one of the larger of many extensive deltaic and outwash deposits that occur close to the limit of marine submergence of coastal regions and in the larger river valleys of Maine. A possible explanation of the abundance of sand at the marine limit is that the ice stood just inland from the sea for some time. Continued glacial erosion under the ice, coupled with the production of meltwater, carried the sand and gravel to the ice-age beach.

Mesozoic Stocks

West of Maine 5 in Shapleigh, Newfield, and Parsonsfield is a line of small Mesozoic intrusions that form, from south to north, Abbott Mountain, Picket Mountain, and Randall Mountain. The Picket Mountain stock is a granite while the other two are syenite, a rock much like granite, but with too little silica to form the usual quartz that is typical of granite. These are part of an irregular line of similar igneous rocks that extends from York, Maine, on the coast, through New Hampshire to Montreal, Québec.

Basalt Dikes in Cornish

Watch just north of the town line of Cornish for an outcrop east of Maine 5. A thick granite dike intruded the Ringemere formation, forming a gneissic rock near the contact. The granite probably came from the nearby Sebago pluton. It is in turn cut by several basalt dikes of Mesozoic age. The country rock southwest of the Sebago pluton consists mostly of schist. Gneiss occurs only close to intrusions.

Basalt dikes of Mesozoic age cutting granite that intrudes the Ringemere formation in another part of the outcrop. This is in Cornish.

Gneissic rocks in the Ringemere formation in Cornish. Gneissic rocks occur only close to the Sebago and other plutons in this area.

Ossipee River

The largest tributary to enter the Saco River in Maine is the Ossipee River, which rises in mountains west of Ossipee Lake in New Hampshire. For many years the Ossipee region has been the source of sand and gravel that is shipped to the Boston Sand and Gravel Company by rail. A great deal of this sand from these deposits is also transported by the Ossipee River to the Saco, where it adds to the sandy bottom of that river.

Hiram Falls and Vicinity

A series of cascades drops 75 feet on the Saco River below the Central Maine Power Company's hydroelectric dam in Hiram. The falls cross the Ringemere formation, which is intruded by many dikes of granite.

Between Hiram and East Brownfield, Maine 5 crosses an unnamed Devonian granodiorite, a somewhat darker rock than granite. In the north part of Hiram an outcrop of this rock encloses a block of gneiss, probably of the Ringemere formation.

Sebago Pluton

Near the intersection of Maine 5 and Maine 160 at East Brownfield, we leave the metamorphosed sedimentary rocks and cross into the

Gneissic rocks of the Ringemere formation intruded by thin veins of granite at Hiram Falls, Hiram.

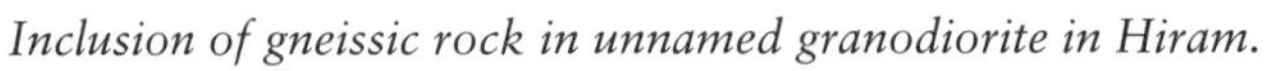

Inclusion of gneissic rock in unnamed granodiorite in Hiram.

Sebago granite. It is the largest continuous pluton in Maine, and it continues under a different name into New Hampshire. It is difficult to explain the origin of this granite in that it is younger than the Acadian collision that formed most granite in Maine and is older than the Alleghanian mountain-building event.

The Sebago granite is a sheet less than 1 mile thick that is inclined down to the north, under the sedimentary rocks there. Hot fluids rising from the cooling granite invaded the overlying rocks and formed the numerous pegmatite bodies of Oxford County.

Lovewell Pond Delta

At what is normally the outlet of Lovewell Pond, flooding of the Saco River has built a delta nearly a mile into the lake. Floods in the Saco River, particularly those from melting snow in the White Mountains, rise quickly and produce water levels that are higher than the lake. This causes a reversal of the normal flow in the outlet stream and forms a current strong enough to transport sand to the delta at the south end of Lovewell Pond. A meandering channel across the delta has prominent natural levees deposited when the water leaves the channel during large floods. Waves rework the sand on the delta front into sandbars, which extend like wings on either side of the delta front.

Pleasant Mountain Pluton

At 2,000 feet, Pleasant Mountain is the highest elevation within the Sebago pluton. Most of the Sebago granite has undergone granular disintegration to such an extent that it can support only low hills and valleys. Pleasant Mountain is buttressed by another of the Mesozoic intrusions found in southwestern Maine; this is a syenite, which is much like a granite although generally darker and lacking free quartz as a major mineral.

Jockey Cap

Just north of U.S. 302, almost in downtown Fryeburg, is Jockey Cap, a nearly bare and rounded height of land that gives a broad view of the lowlands underlain by the Sebago granite and the mountains of Maine to the north and the White Mountains to the west. Jockey Cap is also underlain by Sebago granite.

Old and New Channels of the Saco near Fryeburg

In 1853, the course of the Saco River was shortened by about 10 miles to make the driving of logs easier and faster. The Old Channel, still visible on the ground and maps, went far to the north to near Kezar Lake, causing the river to have a lower gradient, with sluggish flows. Because channels on these low-gradient streams cannot carry water very quickly, water often leaves the channel, causing floods. Log drives used either natural floodwaters or increased flows from the release of water from dams, both known as a driving pitch. When the water left the low-gradient channels, so did the logs. Crews would

Old and new channels of the Saco River.

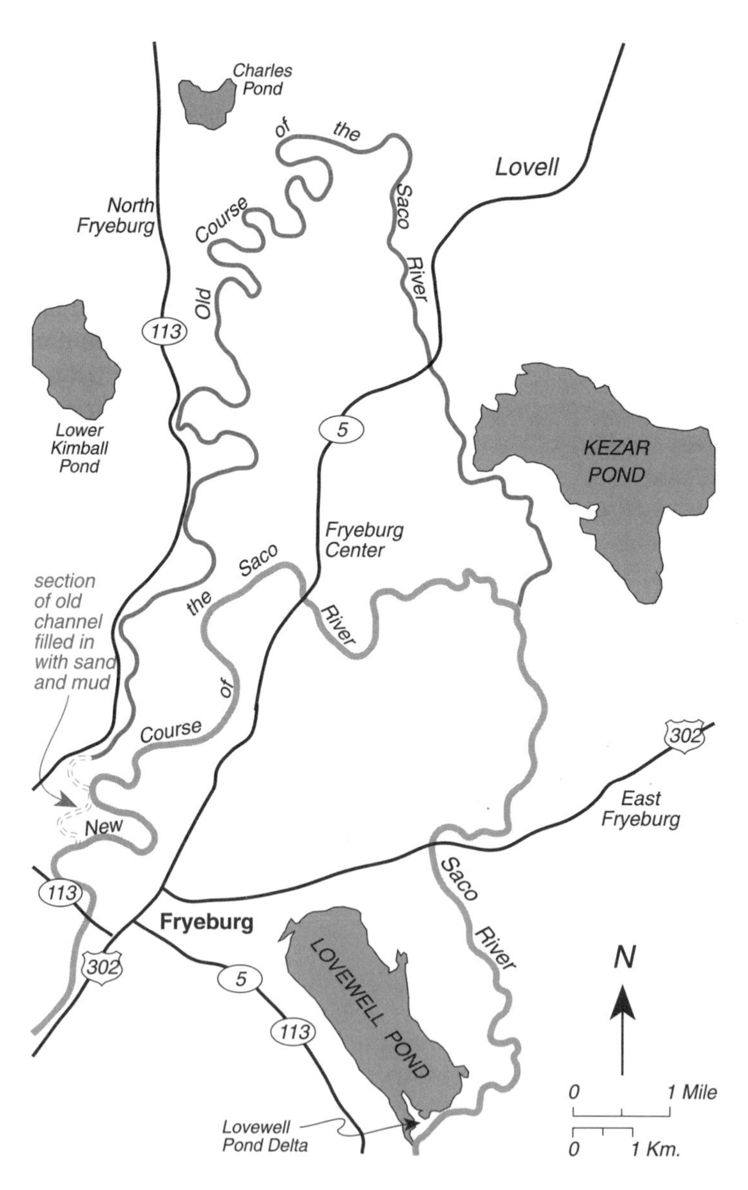

then have to roll, carry, or skid the logs back to the river, usually when the driving pitch was reduced. Another reason for the change was that the Old Channel was more sinuous than the new one, causing logs to jam, especially in the old days when tree-length logs were driven down the Saco.

Most streams in Maine and New England flow downhill from the highlands in the interior toward the coast, generally to the south. In his poem "North Running Brook," Robert Frost compared streams like the Old Channel of the Saco to the cantankerous and independent nature of New Englanders, especially Vermonters.

About 3 miles north of the new-channel crossing near Fryeburg Center, Maine 5 crosses the Old Channel of the Saco. No longer connected with the headwaters of the Saco, most of the flow in the Old Channel comes from the north—from Kezar Lake and the Cold River. The Old Channel was shaped by larger flows than it now experiences. The present flow does not need the whole old channel and has filled in most of it with sand and swamp deposits. The Old Channel of the Saco is now an example of an underfit stream, not big enough for its confines. Most underfit streams form naturally.

Glacial Lake in Center Lovell

During the retreat of the last ice sheet, a glacial lake filled the Kezar Lake basin and Kezar River valley almost to North Waterford. The lake had a boomerang shape. If you wanted to make a lake now as big as Glacial Lake Lovell you would need a dam to hold the water in. It is not known if this dam was ice, perhaps from a large stranded block, maybe a tongue of the glacier coming down the Saco River valley, or some kind of glacial deposit. The extent of the lake is marked by laminated clay and silt washed from the ice into the lake.

Amethyst Crystals on Deer Hill in Stow

Beautiful deep purple amethyst crystals were discovered years ago on Deer Hill in Stow. Other deposits of the same mineral have recently been found in the same area. Amethyst is a form of quartz, with iron oxide possibly providing the color. The Deer Hill amethyst occurs in a somewhat unusual pegmatite for Maine. Beryl, smoky quartz, and pyrite also occur in this deposit.

Songo Pluton

Near North Lovell, Maine 5 crosses into the Songo pluton, a granodiorite to tonalite of Devonian age. A tonalite is like a granite in appearance, but it has a high percentage of plagioclase feldspar and very little orthoclase. This type of rock is also called plagiogranite and trondhjemite. The rocks of the Songo pluton underlie the remaining portion of Maine 5 north to Bethel. At North Lovell bare ledges of this rock are visible on mountains to the north in Stoneham.

Lynchville and Vicinity

The Maine Department of Transportation erected a famous road sign in Lynchville, at the intersection of Maine 5 and 35. The sign lists some of the towns in Maine named for foreign countries and cities. I have a theory that when Maine was part of Massachusetts before 1820, all of the good English town names had been used up by the time the interior of Maine was settled, and these towns had to be called something else. It is only in towns founded after 1820 that place names in Massachusetts are duplicated in Maine. It may also be true that Maine settlers were romantics who dreamed of faraway places, but not having the opportunity to travel they used their town names as a way to satisfy their wanderlust.

For about 3 miles north of Lynchville, Maine 5 and 35 are on an esker. Eskers are ridges of sand and gravel originally formed in a tun-

Road sign in Lynchville at Maine 5 and 35.

nel beneath a melting glacier. The curves in the road follow the meandering path of the tunnel.

Bumpus Pegmatite Mine

A pegmatite is a type of granite with mostly large mineral grains, a few of which are semiprecious gemstones. The Bumpus pegmatite, along with the many others in Oxford County, was formed by magma that escaped from the underlying Sebago pluton and intruded the Songo tonalite. The mine is in Albany Township.

At least fifteen minerals have been identified in the Bumpus mine, but it is noted for its large beryl crystals and rose quartz. In 1928 some large beryl crystals were uncovered. In 1949 a 26-ton beryl crystal was found: It was 27 feet long and 4.5 feet wide at one end and 9 inches wide at the other. The crystal is said to have been found in a mass of deep rose quartz, and reportedly the sight brought tears to the eyes of the miners who first saw it, whether because of its beauty or its value I do not know.

Beryl crystals uncovered in 1928 in the Bumpus mine, Albany Township. —Jane C. Perham photo, Perham's Mineral Store

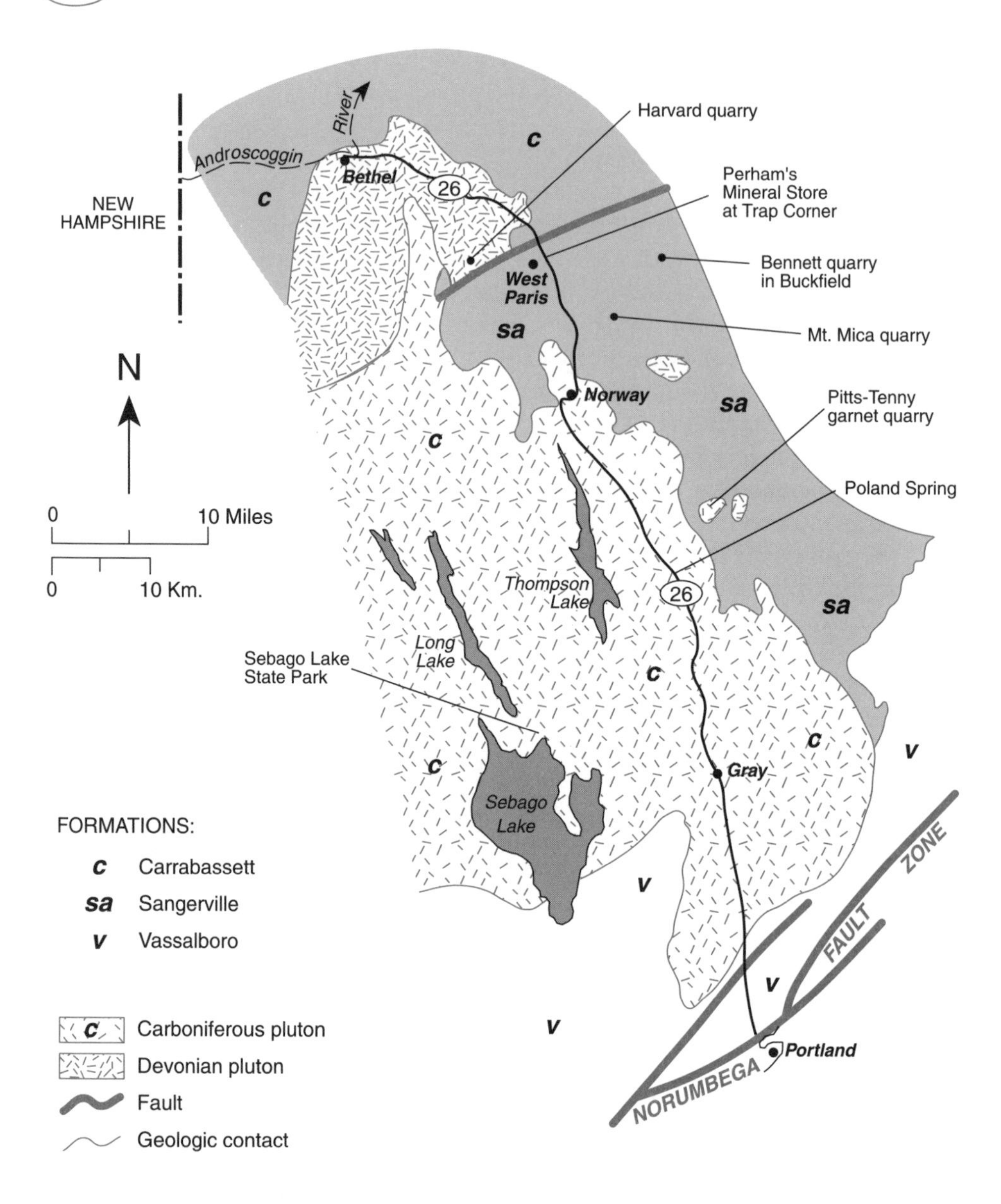

Geologic features along Maine 26 between Portland and Bethel.

Maine 26
Pegmatite Alley
Portland—Bethel

67 miles

Maine 26 runs close to I-495 as far as Gray, then ventures through the heart of the Oxford County pegmatites and into the mountains of northwestern Maine. Between Portland and Gray this road is close to Sebago Lake. Heat from the Sebago pluton affected all the rocks in that area by increasing their level of metamorphism. To the north, the many granite pegmatites were formed from late magma leaving the cooling Sebago pluton.

Glacial marine sediments cover most of the bedrock between Portland and the area just north of Gray, at Sabbathday Lake. They date from about 13,000 years ago, when seawater flooded coastal Maine to a depth of as much as several hundred feet. These deposits reach inland to a line of large deltas, one of which forms the southeast shore of Sebago Lake. Another underlies the highway in the 11 miles north of Gray, and for nearly that distance between Dry Mills and Sabbathday Lake.

Sebago Lake State Park

Sebago Lake State Park is at the north end of Sebago Lake, at the mouth of the Songo River. This river flows through a large outwash plain, which supplied a large quantity of good beach sand to the lakeshore. Other beaches on the west end of the park are made of sand and gravel, also eroded from glacial deposits.

Like many lakes in Maine, Sebago Lake is almost entirely within a granite intrusion, in this case the Sebago pluton. The only part that is not in granite is along the south shore, where you can see schists of the Waterville formation. Glacial deposits line much of the shore.

You can see good outcrops of the Sebago granite between Naples and Witch Cove Beaches on Sebago Lake. The surface of these rocks shows how the mineral grains in the granite, mostly feldspar and quartz, separate as the rock weathers. Ice-age glaciers eroded this loose debris, then left it in their moraines and outwash deposits.

The granite outcrops contain a small vein of pegmatite, obvious because its mineral grains are much larger than those in the normal granite. Some of the smaller grains in the pegmatite are glassy pink and red garnets that remind me how I got started in geology. I spent long summer days on the shores of a lake northeast of here, trying to pry garnets out of granite and schist along the shore. The elegant shape and bright red color of garnets fascinated me, as they do many people.

A basalt dike about 20 feet across cuts through the granite at the east end of Witch Cove Beach. It also cuts several small veins of pegmatite, convincing us that it is younger than either the granite or the pegmatite, which crystallized during Carboniferous time, about 340 million years ago. Like other basalt dikes near coastal Maine, this one formed about 200 million years ago, when Triassic time was phasing into Jurassic time. The basalt magma filled one of the fractures that opened as the Atlantic Ocean first began to open—as North America began to drift away from Europe and Africa.

A number of lakes north and west of Sebago Lake are long and narrow, and appear to branch from it. Their pattern suggests that these lakes lie in former stream valleys that drained through the Sebago area. Glacial erosion probably deepened the basins, including the one that holds Sebago Lake. Some of the lakes, like Long Lake, are so straight and long that they must lie in fault zones. No faults have been mapped in the Sebago pluton, but it is hard to find faults in rocks as homogeneous as granite.

Poland Spring

Water for bottling at Poland Spring flows from fractures in the Sebago granite, where it is collected in large tanks. As with water from many natural springs, special curative powers of the Poland water were claimed, and a large hotel, still standing, housed guests seeking the waters.

Pitts-Tenny Garnet Quarry

The Pitts-Tenny garnet quarry is in Minot off Maine 119, just north of its junction with Woodman Hill Road. The garnets are one of several calcium silicate minerals that occur here. They formed when a mass of molten granite magma invaded limestone in the Sangerville formation during Devonian time, perhaps about 390 million years ago. Steam carried heat and mineral matter from the granite into the limestone, converting it into a mass of calcium silicate and other minerals. Rocks of this kind are called skarns. Miners and mineral collectors love them.

The cinnamon garnet known as grossularite is the most abundant. Like all Maine grossularite garnets, these are brownish red. The first grossular garnets described were green, probably stained by chromium, and were named for their resemblance to gooseberries. Also present

Cinnamon garnets from the Pitts-Tenny quarry in Minot.

in the Minot deposit are diopside, a green pyroxene that contains calcium and magnesium, and a snow-white calcite that fills in the spaces between the other minerals. Complete garnet crystals have twelve crystal faces, but the cinnamon garnets here are only half formed; the rest is buried in a massive growth of garnet, calcite, vesuvianite, and diopside.

Snow Falls

The Little Androscoggin River carved an impressive gorge in schists of the Sangerville formation at Snow Falls in West Paris. It is about 30 feet deep in places and about 300 feet long. Swirling eddies in the flooding river turned pebbles on the bedrock floor of the stream. They drilled cylindrical potholes in the bedrock, which coalesced to form the gorge.

The gorge is eroded into schist laced with small veins of pegmatite. The Sebago granite lies at a shallow depth below these rocks; it provided the pegmatite magma and the heat to metamorphose the rocks. Across Maine 26 from the falls, a striated glacial pavement occurs in this schist.

Oxford County Pegmatites

More than a dozen quarries have worked pegmatites in Oxford County. Many of them are world-famous for the size, the beauty, or

Striated glacial pavement in Sangerville schist near Snow Falls, West Paris.

Pegmatite block in Jane Perham's rock garden at Perham's Mineral Store just east of West Paris. The mineral under the knife is garnet, the black triangles are tourmaline, and the white mineral is feldspar. The flowers are marigolds.

the rarity of the minerals they produced. Look for minerals mainly in the mine dumps, using a small spade or trowel. You need some water to wash the dust off. Finding anything of real value requires hard work, patience, a sharp eye, and luck. Most quarries are privately owned, and you should seek permission to collect specimens.

The Mt. Mica quarry is in Paris, about a mile east of Paris Hill—one of the most beautiful small villages in Maine, with a marvelous view of the White Mountains to the west. More than fifty minerals have been identified from this quarry. It is most famous for the large crystals of gem quality tourmaline that have been found in shades of green, pink, blue, and white, and in combinations. The largest specimens with the best crystal forms line the sides of open cavities, called vugs. Some are large enough for a person to crawl into. As the pegmatite fluids entered these openings, the crystals could grow into their natural forms without interference from neighbors.

The Harvard quarry is about 5 miles west of West Paris. In 1923 and 1924 it supplied Harvard University with many choice specimens for its mineral collection. Most of the usual Oxford County pegmatite minerals have been found in it, as well as purple and blue apatite, cassiterite, spodumene, and lepidolite. Few collectors climb the steep hillside to reach it.

The Bennett quarry is in Buckfield on the Paris Hill Road. Included among the more than forty minerals found in this quarry are blue, green, and pink beryl as well as tourmaline in the same colors. Unusually large crystals of rose beryl were found in the Bennett quarry in 1989, one about 3 inches wide and the other a foot across. These crystals turned pink upon exposure to sunlight. Pink beryl is known as morganite. Also present are the lithium minerals lepidolite and spodumene.

Disrupted Streams

The town of Bryant Pond is named for a pond that is part of the Little Androscoggin River drainage, which begins a few miles north of here with Meadow Brook. Little Androscoggin River flows south to join the main Androscoggin River downstream of Auburn. Meadow Brook shares its wide valley with another stream, Barkers Brook, which flows north and enters the Androscoggin near Rumford. No one knows for sure why two brooks flow in opposite directions in the same valley with nothing much separating them.

North of the Androscoggin River between Bethel and Rumford are long tributaries that originate in the mountains west of Rangeley. Opposite each is a wide valley containing a small north-flowing brook, and then a much longer one flowing southward. It is possible that at some time in the past all the streams were parts of continuous rivers that flowed south. The Androscoggin River then cut eastward across this system, stranding the north end from the south end of each river.

The last glacier knew about these channels. In several places eskers and other meltwater deposits begin in the northern tributaries, cross the Androscoggin River, and continue down the southern valleys. One of the most dramatic of these deposits is an esker in the Ellis River valley north of the Androscoggin River that crosses into the Barkers Brook valley, easily mounts the tiny climb up and over to run beside Meadow Brook, and continues down the Little Androscoggin River.

North and South Ponds

Maine 26 between Bryant Pond and Lynchville crosses between two ponds, one north and one south of the highway, apparently giving the ponds their names. Part of this road that separates the ponds is on an esker. Visible from the road in North Pond is a good example of glacially streamlined bedrock, here granodiorite of the Songo pluton.

Interstate 95 and 295
Brunswick—Bangor

107 miles

These sections of I-95 and I-295 cross the rusty-weathering Cushing formation, the calcareous mudstones of the Vassalboro formation, and the silvery slate of the Waterville formation. The Cushing formation was laid down during late Precambrian or early Paleozoic time; the Waterville formation dates from Silurian time; the Vassalboro formation was deposited during Ordovician and Silurian time. Much of the route follows the trends of the layering, so you travel a long way on the same rock.

Roadcuts along the interstate expose at least two of the abandoned quarries in the Topsham pegmatite district. The famous granite quarries of Hallowell, near Augusta, supplied stone for monuments and buildings along the East Coast and in the Midwest. These segments of I-95 and I-295 also cross glacial marine deposits left as the last ice sheet in this part of Maine finally melted.

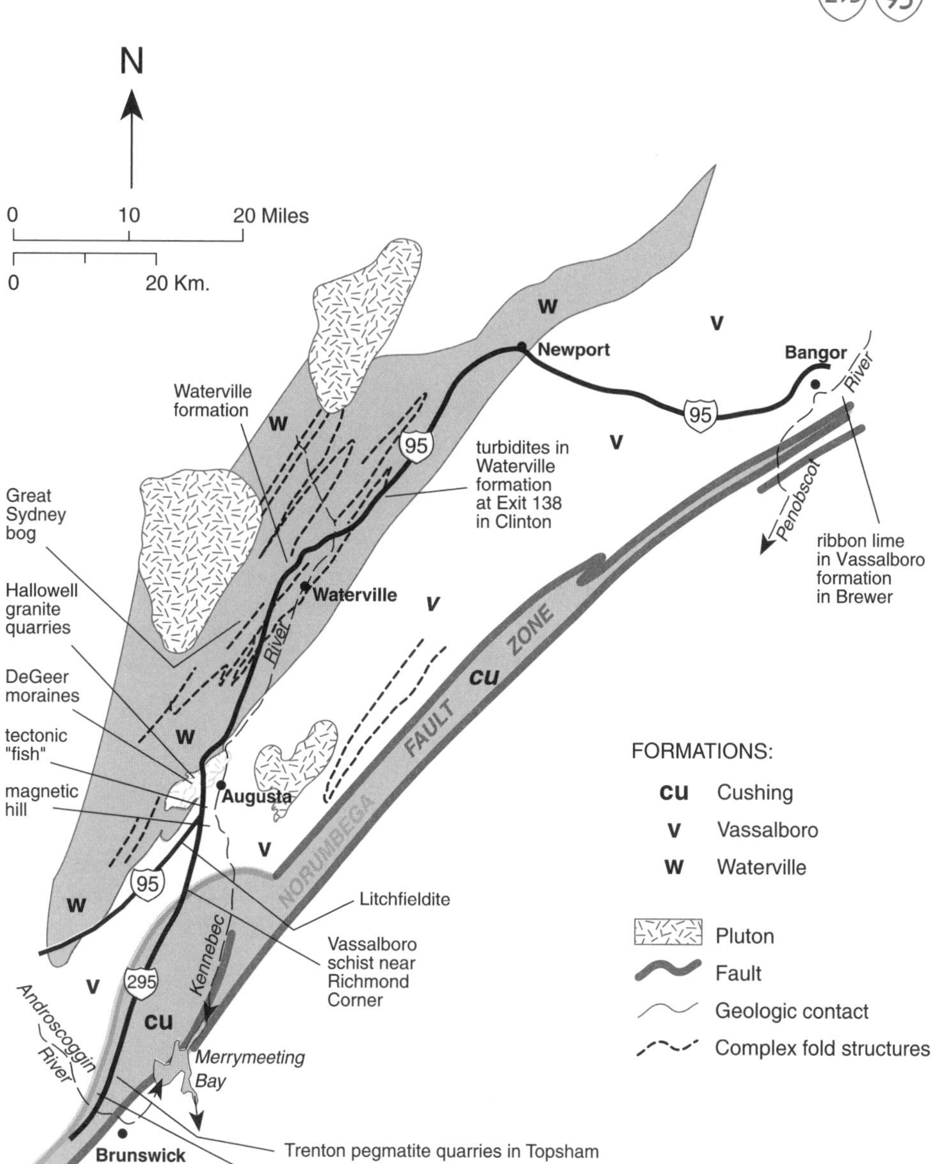

Geologic features along I-95 between Brunswick and Bangor.

Cushing formation exposed on I-295 in Brunswick. The horizontal layers in this outcrop are rare in Maine, where most sedimentary layers are steeply tilted.

The Kennebec and Androscoggin Rivers, the second and third largest in Maine, meet in Merrymeeting Bay near Brunswick. Merrymeeting Bay is a stopping place for migrating geese and ducks that follow the Atlantic flyway. The Androscoggin River was one of the most polluted streams in the country during the first half of the twentieth century, when paper mills in Berlin, New Hampshire, and in Rumford and Livermore, Maine, dumped their effluents into it. Partly as a result of the federal Clean Water Act, the river is no longer known as "the Androstinking."

High-Water Marks in Topsham

Look north of the highway between the Androscoggin River and exit 31 for an old beach that dates from the time when seawater flooded coastal Maine as the last ice age was ending. It is likely that the sea rose even higher, submerging the hill behind the beach. Then, as the land rose, sea level fell to this position for a few years while the waves reworked glacial meltwater deposits into this beach. Several other beaches were built later, as sea level kept falling. The wind blew some of the sand into dunes. The land is now kept clear for blueberries, but the sand shows through in a few places.

Pegmatite of the Topsham Area

Interstate 95 cuts through two old pegmatite quarries. Watch on the east side of the northbound lane between mileposts 78 and 79 for two deep and narrow trenches cutting through the rusty beds of the Cushing formation. These were part of the Trenton pegmatite quarries, where mining began in 1860. In the early years, feldspar was shipped down the Cathance River to the Kennebec River, where it was loaded on ships and carried to ceramic plants near Trenton, New Jersey. Later, all the feldspar from this district was milled in Topsham and shipped by rail. The last shipments went to East Liverpool, Ohio. These quarries ceased to supply feldspar to the ceramic industry when other mines began to produce a better product.

Some quarries briefly produced feldspar for bird grits during the 1960s. Lacking teeth, birds grind their food in a pouch lined with strong muscles, the gizzard. Along with their food, birds pick up stones to help grind it. The mine shipped grits in sizes for turkeys, chickens, and parakeets.

Vassalboro schist with a vein of pegmatite at exit 43, Richmond.

Abandoned feldspar quarry near milepost 35 in Bowdoin. Under the pine tree above the quarry is some feldspar waiting for shipment.

Magnetic Hill

Just south of Gardiner on U.S. 201 is a deposit of magnetic iron sulfides large enough to deflect a compass by many degrees. The magnetic mineral is pyrrhotite, part of the Cushing formation that melted in the heat of a nearby granite intrusion. The molten magma cooled to make a rock rich in iron and sulfur. Pyrrhotite commonly contains nickel, but the Gardiner deposits do not contain enough to mine. This deposit is similar to the one at the Katahdin Iron Works between Brownville and Millinocket, whose weathered upper surface was mined for iron ore in the 1800s.

Litchfieldite

Few towns anywhere that I can think of have rocks named for them, although a few are named for rocks. The town of Litchfield does. Litchfieldite is one of a group of plutonic rocks described as silica deficient, which means that they contain too little silica—silicon dioxide—to make the minerals that would normally form granite and similar rocks.

Two minerals occur together in litchfieldite that would have been feldspars had more silica been available: sodalite, which is blue, and cancrinite, which is yellow. The town hall in Litchfield has on display a large specimen of this unusual and colorful rock. You can collect litchfieldite from outcrops along Maine 126 and 9, about 5 miles west of where they cross I-295 near the tollgate. Outcrops of litchfieldite occur on both sides of the road.

Tectonic Fish

Watch for the curious outcrops of the Vassalboro formation east of I-95, just south of the Maple Street overpass, near milepost 107. They contain nodules of calcium carbonate that resemble jumping fish. The bedding in this outcrop is rather simple, and the fish are parallel to the bedding. These may have started out as calcium carbonate concretions that were squeezed into the fish shapes during folding.

Hallowell Moraines

Just south of the Augusta exits, near milepost 109, I-95 crosses numerous moraines deposited when the glacier of the last ice age was grounded on the seafloor. These deposits, DeGeer moraines, form small hills that are rounded and rather elongate.

Tectonic "fish" in Vassalboro formation near milepost 107 in Farmingdale. This rock is rich in calcium carbonate.

DeGeer was a Danish baron who studied glacial features around the Baltic Sea. He concluded that moraines like these in Maine record annual positions of the retreating ice margin. People who have studied them around Hallowell and in the rest of Maine have found no evidence that would either substantiate or discredit that idea.

Moraines on both sides of the nearby Kennebec River make a chevron pattern that points upstream. It suggests that a reentrant in the ice margin extended up the river. That happened because the water in the middle of the valley was too deep for the ice to touch bottom until it extended farther upstream. An ice margin on land typically reaches down valleys because the ice is thicker there.

The moraines consist of till, with sand and gravel on their south sides. The whole deposit is coated with glacial marine mud. The glacier sat on the seafloor on the north (upstream) side of these bumps in several hundred feet of water. The moraines are about 13,000 years old.

Hallowell Granite Quarries

The Hallowell granite is Devonian in age, as is most granite in Maine. The quarries are west of the interstate, close to the Augusta exit at milepost 110. They have been closed for years. Waste granite blocks from one of them scatter down the slopes to within 300 feet of the highway. The Hallowell stone is pale gray—white granite in the building trade.

DeGeer moraines in Hallowell on the west side of I-95.

Waste piles of granite from Hallowell quarries, milepost 110 in Hallowell.

Oxen hauled granite blocks down Winthrop Street to shops in Hallowell, where it was prepared for shipping down the Kennebec River. Carts coming down the steep grade into town were restrained by snubbers, ropes or cables wrapped around trees or snubbing posts, to prevent the cart from running away and crushing the oxen or perhaps running right into the river.

In their time, the Hallowell quarries supplied granite for such notable buildings as the state capitol in Albany, New York, the post office building in Houston, and the library at the United States Naval Academy. The capital of Maine was moved from Portland to Augusta, where the new capitol was built of Hallowell granite between 1829 and 1832.

Augusta Airport Delta

The Augusta airport, like many in New England, is on a delta, in this case a glacial marine delta. The flat upper surfaces of deltas make excellent sites for runways.

The Augusta delta is at the southern end of a long series of eskers that extend nearly 100 miles up the Kennebec River. They are the remains of meltwater streams that flowed within the ice, carrying sand and gravel to the delta. Bedding within the delta records the stand of sea level when it was deposited.

Holes dug in deltas invariably reveal a pattern of inclined layers, with flat layers on top of them, what geologists call foreset and topset

beds. Sea level was where the flat layers intersect the inclined ones, a few feet below the top of the delta, now about 340 feet above sea level.

Elevations of other deltas rise about 2 feet per mile toward the northwest. That records the rise of the crust as it floated up after the load of glacial ice melted. Deltas in glacial lakes in other parts of New England generally show a slope of about 4 feet per mile. The lower slope in coastal Maine reflects the simultaneous rise of sea level, world-wide, as the ice melted.

The Great Sidney Bog

Many of the lakes in this area stretch out to the northeast, in the direction of the folded beds of Silurian mudstone in the Sangerville formation. The lakes are generally on limestone beds within this unit. Messalonskee Lake in Sidney and Belgrade is about 12 miles long and 1 or 2 miles wide at its south end. The lake was bigger in the past, soon after the last glacier melted, but became at least 2 miles shorter since filling with peat. Such deposits are inspiring new interest for use in specially designed electrical generating plants.

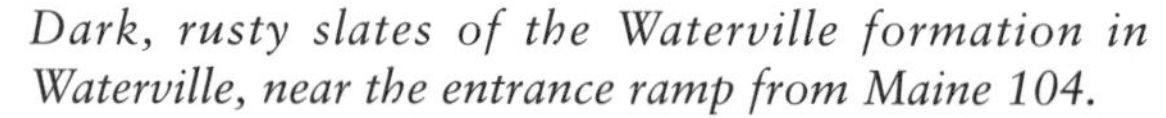
Dark, rusty slates of the Waterville formation in Waterville, near the entrance ramp from Maine 104.

Shiny Rocks along Interstate 95

Interstate 95 passes the best exposures of rocks in the eastern part of the Central Maine slate belt. The gray outcrops between Augusta and Newport have a soft shiny luster. These rocks, most of them in the Waterville formation, have been metamorphosed beyond slate. They are phyllites. In Waterville, some exposures of the rocks are dark with rust.

Phyllites are composed mostly of small crystals of mica that crystallize into flat plates that easily split into thin flakes. These rocks were folded as they recrystallized, so their mineral grains are aligned parallel to each other. Mica flakes in metamorphic rocks invariably line up with their flat surfaces perpendicular to the direction of compression. The rock tends to split along the flat surfaces of the aligned mica flakes, so what you see when you look at those surfaces is millions of minute crystals of mica, all glittering in the light. That is what makes the rock shiny.

A bit more metamorphism would have grown those mica crystals big enough that you could actually see them one by one, instead of just their collective shine. At that stage, the rock becomes a schist. So

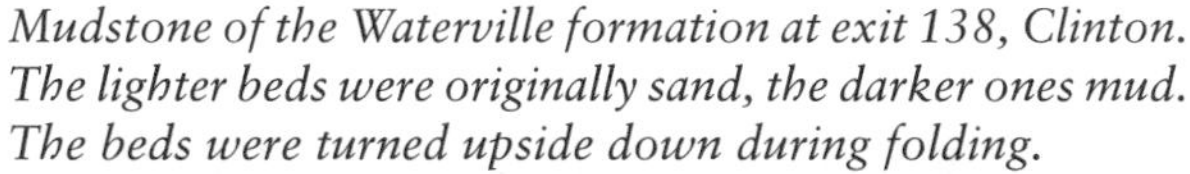
Mudstone of the Waterville formation at exit 138, Clinton. The lighter beds were originally sand, the darker ones mud. The beds were turned upside down during folding.

a little bit of metamorphism converts a mudstone into slate, a bit more makes it into phyllite, still more creates a schist.

White mica, the kind you most commonly see in phyllite and schist, has almost exactly the same composition as ordinary clay. That explains why mica grows in recrystallizing mudstones. Black mica contains iron; it forms during metamorphism of rocks that contain iron oxide as well as clay. An abundance of either kind of mica in a metamorphic rock means that it probably began its career as some sort of mudstone.

Shiny phyllites are much more common in the Waterville formation than in the Vassalboro formation, which is exposed between Portland and Topsham, and between Newport and Howland. It consists largely of calcareous sandstones that do not contain much of the chemical raw material for mica of any kind. They weather to a dull brown color.

Folded, Folded, and Refolded

Watch all along I-95 for squiggles and warped patterns in the roadcuts. These are folds; they date from the Acadian mountain-building event of Devonian time. Some of the best folds are in the Waterville formation, near Waterville. Other nice folds appear in outcrops south of Augusta.

Geologists who study these squashed rocks believe they were folded three times, on three separate occasions. The first event produced recumbent folds, which lie flat. If you sit on the floor and touch your toes, your body is in a recumbent fold. The recumbent folds were later buckled into upright folds in two directions. If you touch your toes while standing up, you are in an upright fold position. The upright folds make the dominant structures you now see. They trend northeast.

Fold a piece of paper twice, one recumbent event and then one upright event. Notice the amount of horizontal shortening: you can easily bend a sheet of paper 11 inches wide into a fold belt less than 1 inch wide. Folding shortens rocks, too.

The amount of folding and horizontal shortening of a blanket of sedimentary rocks tells something about the interactions between the sedimentary rocks and the more rigid basement rocks on which they lie. If you lay a tablecloth on a table, then ask two people to push the table from both ends, that would not fold the tablecloth. Similarly, if some force squeezes both the basement and the sedimentary rocks that cover it, the rigid basement rocks will inhibit folding of the cover. But you can easily fold the tablecloth if you let it slide across the table.

Imagine sediments that originally blanketed a basin floor in horizontal layers. They now make a belt of folded rocks some 100 to 150 miles wide. How wide was the original blanket of sediments? Like the tablecloth, they must have detached and moved over the basement on which they were deposited. Such extreme horizontal shortening commonly occurs as sediments are scraped off a sinking slab of oceanic crust and into the oceanic trench. How wide was the original blanket of sediment? As wide as the floor of the Iapetus Ocean.

Ups and Downs of the Interstate

Unlike most of I-95, the stretch between Newport and Bangor traverses the rock formations across the trend of the bedding. Along the rest of the interstate, we see the same rock strata for many miles. In this section, the road crosses different types of rock. In the 25 miles between Newport and Bangor, I-95 crosses first low swampy ground, then a small hill, and it repeats this many times. Each hill has several outcrops of the Vassalboro formation. There are few rocks showing in the lowland, probably covered with 10 feet or more of glacial and

Ribbon lime of the Vassalboro formation on I-395 in Brewer.

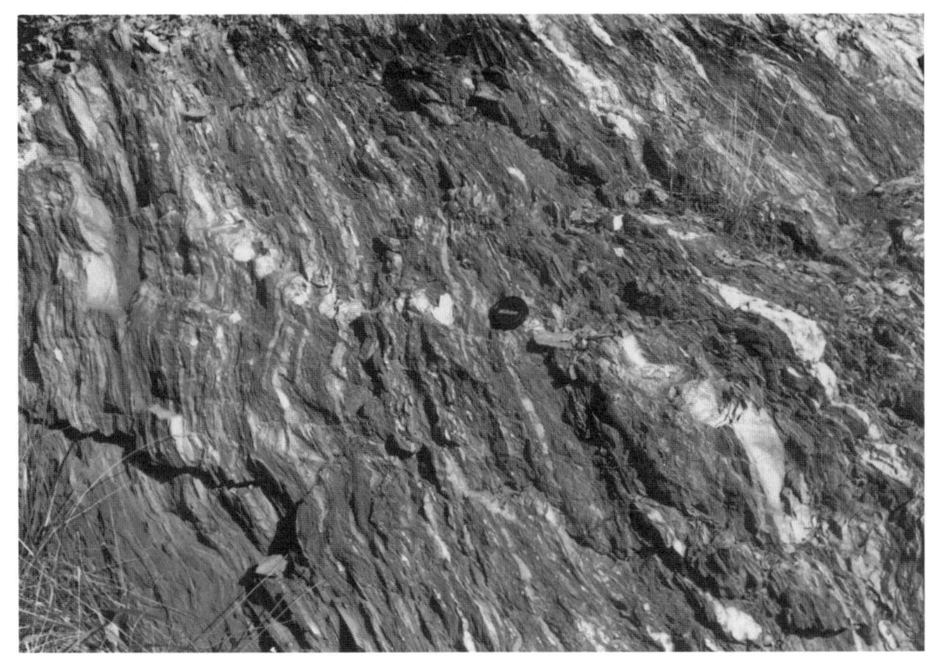

swamp deposits. Within the Vassalboro formation, patches of stronger rock alternate with weak ones. The stronger beds have more sandstone, the weaker ones more shale and limestone. Long before the glaciers came, the lowlands were etched out of these soft rocks through weathering and erosion by streams. Glaciers may have dug a little deeper into those fragile rocks before covering them with till. The more resistant beds form the small hills.

Interstate 95
Portland—Augusta

60 miles

Outcrops of Sebago granite line the road between Portland and Lewiston. It is about 340 million years old. From Lewiston north, the route crosses mudstones deposited during Silurian time. Most of the outcrops also contain some limestone.

Presumpscot Formation

The locality for which the Presumpscot formation was named is near I-95, along the Presumpscot River. It contains the fossil shells of sea animals about 13,500 years old. It is the deposit of glacial marine mud that covers so much of the coastal area of southern Maine.

Sheeting in the Sebago Granite

Sheeting in granite occurs as the rock expands during long periods of erosion. When the granite was more deeply buried, it was compressed under the weight of overlying rock. As the rocks and weight above were gradually removed, the granite expanded and the sheeting fractures formed. People who quarry granite take advantage of the sheeting in removing the rock from the earth.

The Gray Delta

The Gray delta is probably the largest in New England. It lies near the inner limit of seawater submergence in this part of Maine. A series of eskers near Sabbathday Lake on the west edge of the delta records the approximate position of the ice margin while the delta was deposited. The delta built out into the sea to the south, east, and northeast, providing a nearly level route for the interstate for about 11 miles.

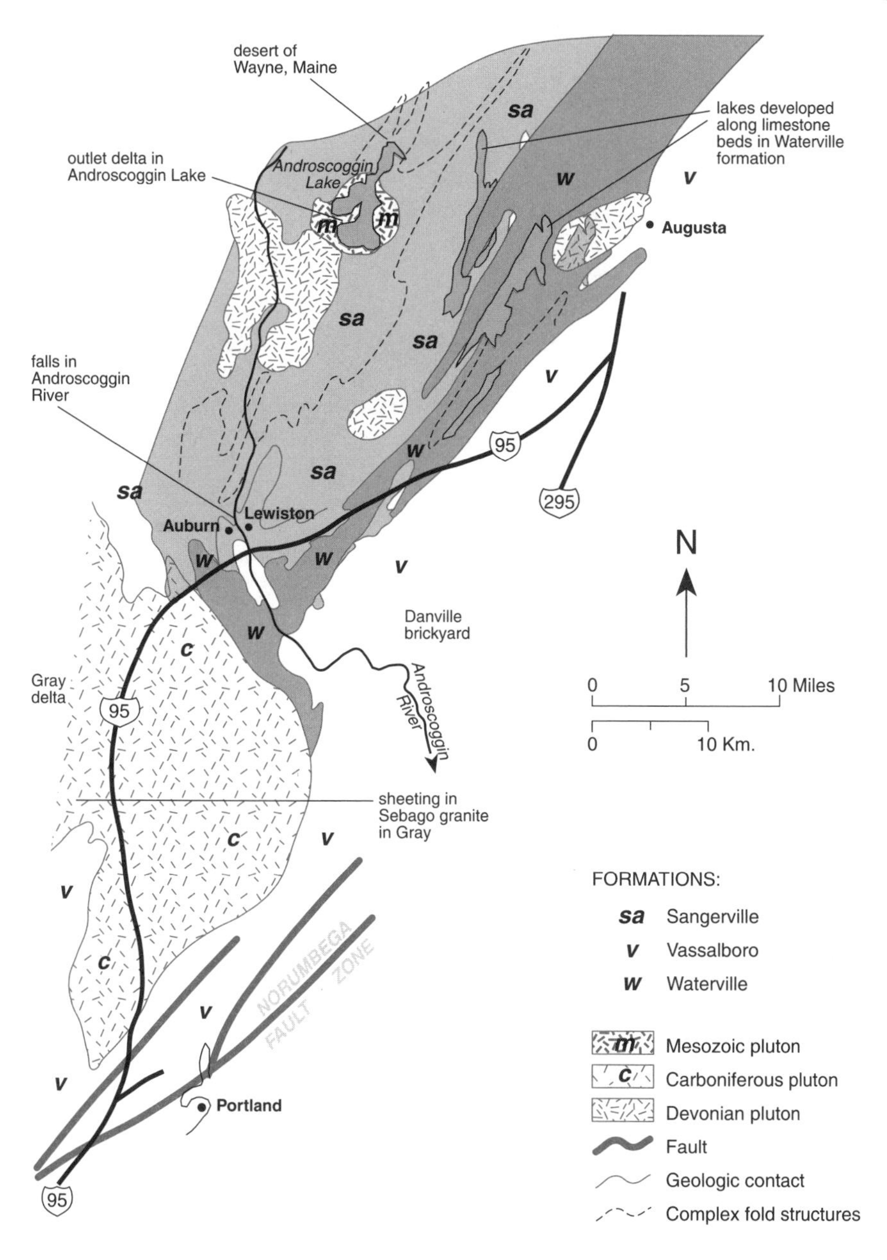

Geologic features along I-95 between Portland and Augusta.

Sheeting structure in the Sebago granite at Gray.

At the south end of the delta, near the New Gloucester tollgate, I-95 crosses the delta's last deposits, the front itself. The size of this delta leads me to suppose that the ice margin stood near here for tens of years, while its meltwater carried about a half-cubic-mile of sand and gravel to the Gray delta. Highway borings near the delta front reveal that its sand deposits built over the silt and clay of the Presumpscot formation. The well-drained soil of the delta supports a growth of red and white pine in a landscape that resembles the coastal plain of North Carolina.

Groundwater feeds the streams that dissect the delta surface. The impermeable clays of the Presumpscot formation that lie beneath the delta trap the groundwater, which finally leaks out through a series of springs at its lower end. The escaping water erodes a channel and deepens it headward, a process called spring sapping. One of these springs is developed for the state fish hatchery at Gray, which raises salmon and trout.

Danville Brickyard

The brick plants at Gorham near Portland and in the Danville section of Auburn are the only two in regular operation in the state, the number down from more than one hundred in 1900. Both produce bricks from the Presumpscot glacial marine mud. They mix the clay with a small amount of sand and other additives to give the bricks the desired surface texture and color. Then the mixture is extruded like toothpaste through wires that cut it into bricks. The bricks are dried and passed on special cars through a furnace, where they bake at a temperature of nearly 2,000 degrees Fahrenheit.

Lewiston and Auburn

Lewiston, the second-largest city in Maine, has been a textile center since about 1820. Most of its people are of French-Canadian descent. Auburn, the fourth-largest city, was for many years a place that sold shoes and the machinery for making shoes. It has lost much of its business to foreign companies. If Lewiston and Auburn were to join to form a single city, as Dover and Foxcroft to the northeast did, it would be slightly larger than Portland, Maine's largest city.

A large waterfall in the Androscoggin River separates Lewiston and Auburn. Years ago it slowed exploration and settlement up the river, while the valleys of the Kennebec and Penobscot Rivers, which lack major falls near their mouths, were explored and settled. The rock that supports the falls is a granite pegmatite that intruded mudstones. This is one of a number of pegmatites, including the famous one at Mt. Apatite in Auburn, that follow a trend northwest. They crystallized from magma that rose from the Sebago granite pluton. Minerals in this pegmatite include feldspar, quartz, muscovite mica, beryl, and black tourmaline.

The falls stop in dry weather because a series of canals diverts the Androscoggin River to mills. The best time to see them is in March and April, when the river runs high from melting snows in the White Mountains of New Hampshire and the Blue Mountains of Maine.

The Desert of Wayne, Maine

The largest area of active windblown sand in Maine is on Berry Road, in the north part of Wayne and extending into Fayette. The total area of dunes, some active, some stabilized beneath grass, is about 200 acres. Strong glacial winds blew the sand from glacial outwash deposits in the Androscoggin River valley to the west before vegetation covered them. The sand passed over the crest of a hill, blasting and polishing outcrops and boulders along the way, then accumulated on its sheltered lee side.

Androscoggin Lake

Androscoggin Lake is also in Wayne, and its southwest corner extends into Leeds. The basin of Androscoggin Lake is eroded in a cir-

Sand dunes in desert of Wayne.

cular mass of dark gabbro, the coarsely crystalline version of basalt. The question of the gabbro's age has inspired some difference of opinion; some geologists think it intruded during Carboniferous time; others prefer to think it intruded during Cretaceous time. Oversized crystals of black pyroxene and pale greenish plagioclase make an unusual pegmatite along the western margin of the gabbro. Beaches on bedrock islands within the lake have high concentrations of magnetite, giving them a shiny black color. If you drag a magnet through that sand, it will come out covered with little grains of magnetite.

Most of the year the Dead River drains Androscoggin Lake into the Androscoggin River, about 6 miles to the northwest. Spring snowmelt raises the Androscoggin River above the level of the lake and reverses the flow of the Dead River. Sediment carried in these reversing flows has built a delta about 2 miles into the lake. The delta stands on the marine muds of the Presumpscot formation. Evidently, seawater flooded the basin of Androscoggin Lake as the last ice age was ending. Years ago, an ingenious flip-flop dam controlled the level of Androscoggin Lake when it was higher than the river and prevented the flooding river from flowing into the lake.

Skinny Lakes in Readfield, Winthrop, and Gardiner

In Winthrop and surrounding towns, a connected series of long, spindly lakes occupies basins in weak limestone units within the

Waterville formation. The mudstones of the Waterville formation form moderate to high hills between the lakes. The limestone units snake back and forth in plunging folds, giving the lakes formed along them a forked appearance. Maranocook Lake divides around Tallwood Point in Readfield. Lake Annabessacook occupies the western fork of a basin that extends into Narrows Pond. This system drains into Cobbosseecontee Lake, the widest in this area because the limestone unit in which its basin lies is thicker than those to the west. These waters finally drain into the Kennebec River at Gardiner, but not before passing through Pleasant Pond, the narrowest of all of these narrow lakes. Pleasant Pond is eroded along a fault between the Cushing and Vassalboro formations.

Interstate 95
Bangor—Houlton

120 miles

In the 120 miles between Bangor and Houlton, I-95 crosses only four formations, and the landscape is monotonous. The highway roughly follows the strike of the fold belt, which explains the lack of variety in rocks and scenery. Side roads cross the fold belt and may cross exposures of two or more formations within a short distance. It is only north of the Penobscot River watershed that things begin to change. Most of the changes occur within sight of the highway.

U.S. 2A, which parallels I-95 between Bangor and Houlton, follows part of the Military Road built to get troops to the Aroostook War from 1839 to 1842. American and British troops marched around Aroostook County in that border dispute without firing a shot.

Vassalboro Formation

Watch near Bangor for several good roadcuts in the Vassalboro formation. The typical Vassalboro rocks are interbedded mud and sand that form repetitive light-colored (sand) and dark-colored (mud) bands. These are especially noticeable where the interstate crosses Kenduskeag Stream, which has also exposed the rocks within its gorge.

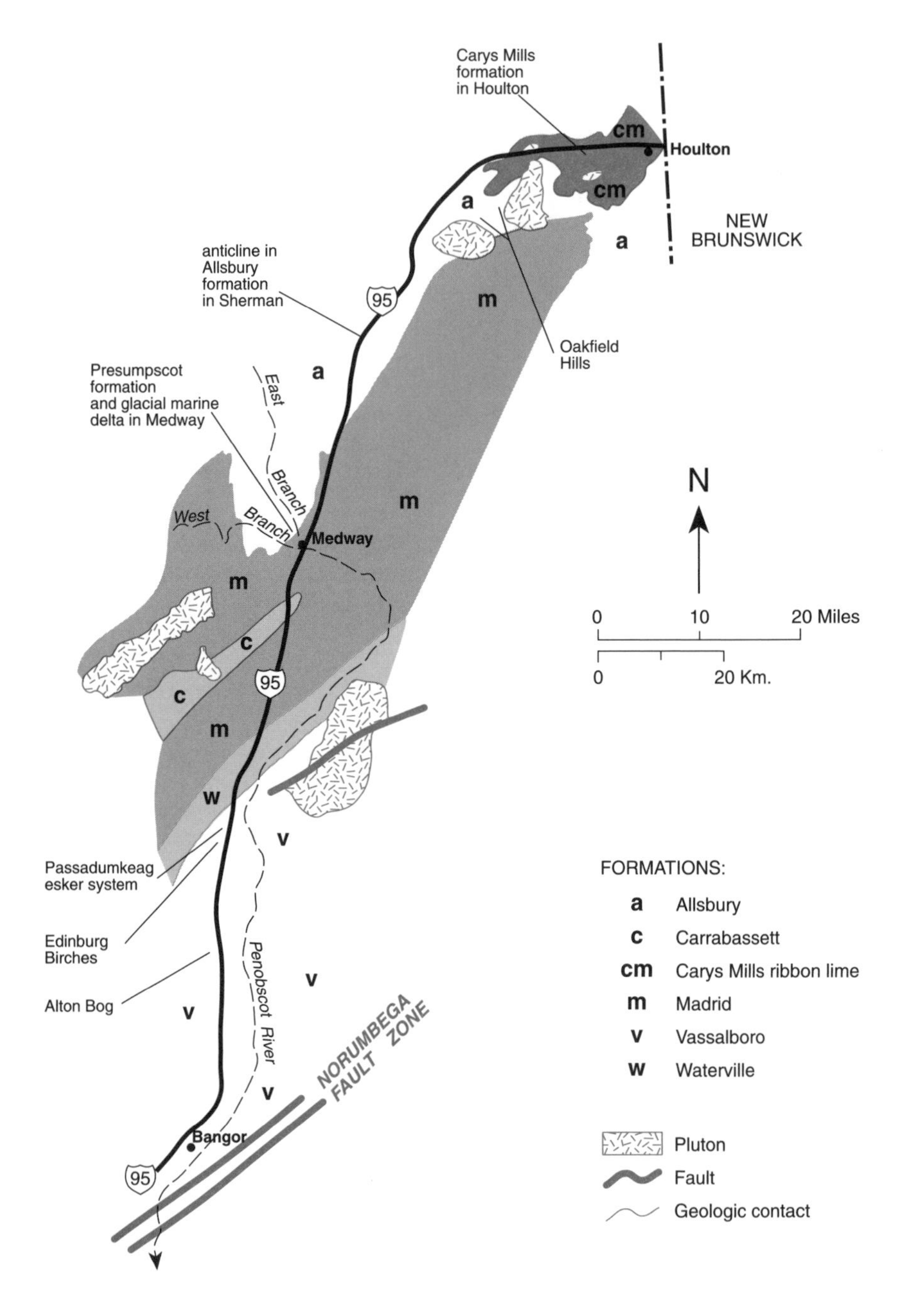

Geologic features along I-95 between Bangor and Houlton.

Old Town, Indian Island, and the University of Maine

Above Old Town, the Penobscot River splits into several channels that flow around three large islands. Old Town and the campus of the University of Maine at Orono occupy the largest and southernmost island. The waterfalls that cascade across bedrock on both sides of this island have long been used for industrial power. The Penobscot Indian Nation has its principal headquarters on Indian Island, close to Old Town. On each of his three trips up the Penobscot River between 1846 and 1857, Henry David Thoreau stopped at Indian Island to engage a guide, although on his first trip they were all too busy hunting moose to accompany him.

Alton Bog

At milepost 200 in Alton, I-95 crosses one of the largest peat bogs in Maine. It is about 3 miles long and 2 miles wide. Sphagnum moss covers most of its spongy surface, and insect-eating plants such as the sundew and the pitcher plant are plentiful. The white cottony blossoms of spring and summer are mare's tail. American larch trees, or tamarack, grow on a few islands of glacial till and creep onto the edges of the bog. Their needles turn a beautiful golden color in the fall, then drop. As soil eroded from the uplands is deposited on the bog, trees will eventually cover it.

Alton Bog began as a pond after the last glacier melted from this area. Swamp vegetation and leaves blown in accumulated on its floor, where no oxygen was available to break down the organic matter. This material has nearly filled the original pond, leaving little open water. Birch Stream flows around the edge of the bog, carrying away the black swampwater full of the organic acids that typically drain from bogs. The forty-fifth parallel, halfway between the equator and the north pole, passes through the bog.

Passadumkeag Eskers

The esker ridges on both sides of the Penobscot River near Passadumkeag are classics, exceptional for their height and length. Topographic maps of the Passadumkeag area appear in elementary geology texts and laboratory manuals.

In Maine, eskers are commonly called horsebacks or whalebacks. The Enfield Horseback, Gould's Ridge, southwest of Cold Stream Pond, rises as much as 60 feet above the surrounding swamps and bogs. The interstate crosses the Hoytville Horseback near Pollard Brook, at milepost 214. A good deal of it is now in the highway subgrade and in the blacktop. Eskers are excellent sources of construction aggregate, so it is common to find them mined. The sand

and gravel in these eskers came partly from the Katahdin granite to the northwest.

Northernmost Presumpscot Formation

Pits in Medway, where the East Branch and the West Branch join to form the Penobscot River, expose some of the northernmost deposits of the Presumpscot formation. The flooding ocean waters drove the glacier up the Penobscot River to Medway. An esker that was under the ice was exposed as the glacier retreated north of Medway. The esker ridge was then buried under mud and sand carried into the ocean from the melting glacier. The crust began to rise as it was relieved of the great load of the ice, the sea retreated, and marine deposition could get no farther inland.

Portion of the topographic map showing I-95 as it crosses the Hoytville Horseback at milepost 214, Edinburg.

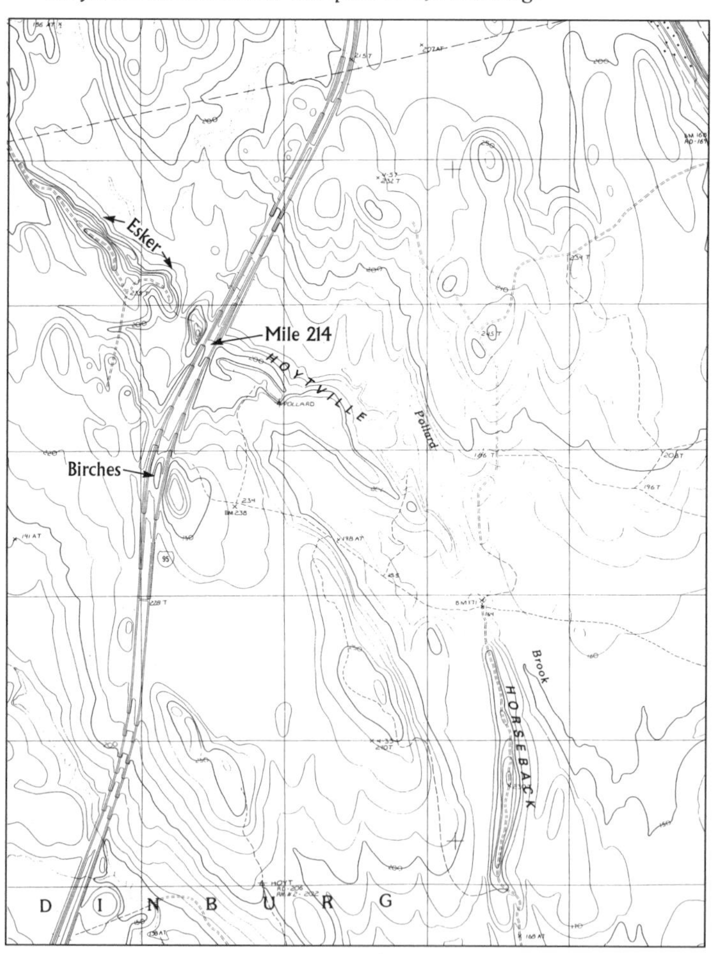

Presumpscot formation mud and sand in Medway. The gravel at the bottom of the pit is a portion of an esker buried by the marine sediments.

A View of Katahdin

North of the Medway exit, the highway crosses Salmon Stream, near a roadside turnout with an excellent view of Mt. Katahdin, about 20 miles to the northwest. You can see the gentle north slope of the summit tableland, which almost exactly follows the top of the Katahdin granite pluton. Bedrock in several lower peaks north of Katahdin is the Traveler rhyolite, which erupted from the magma chamber that crystallized into the Katahdin granite.

You can also see glacial cirques, great hollows on the east side of the mountain, and the redoubtable Knife Edge, along which runs the most morally invigorating trail in Maine. Silurian mudstones of the Allsbury formation lie beneath the low area between the overlook and the great mountain. The low hills in the middle distance are eroded in Ordovician volcanic rocks and the cherts and sandstones associated with them.

Allsbury Formation

From the Medway exit northward for more than 40 miles the roadcuts expose a brown slate called the Allsbury formation. It is Silurian in age and noted for its fossil graptolites, an extinct floating organism usually preserved in rocks as carbonized films that show something of the shape of the organism. They look a bit like hacksaw blades etched on the bedding surfaces.

Allsbury formation exposed at I-95 exit 264 in Sherman.

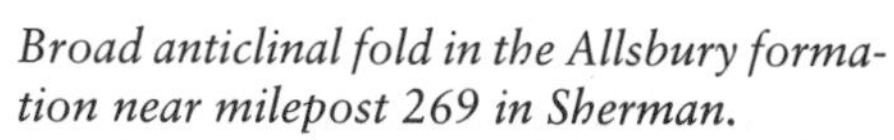

Broad anticlinal fold in the Allsbury formation near milepost 269 in Sherman.

On the east side of the interstate, near milepost 262 in Sherman, is a good exposure of a gentle anticlinal fold in the Allsbury formation. Faults complicate the structure, particularly on the north end of the outcrop.

Taconic Volcanic Chain

A row of conical hills rises on the northwestern skyline between the Sherman Mills and Island Falls exits from I-95. They are eroded in volcanic rocks that erupted during Ordovician time. Never mind the conical shapes, these are not the original volcanoes, just hills eroded in volcanic rocks. No volcanic landform could survive the 440 million years since Ordovician time.

The rocks in those hills erupted in a chain of volcanic islands that stood off the early Paleozoic coastline of North America. It probably resembled the modern Aleutian volcanic chain, and, like it, formed above a slab of oceanic crust sinking into the oceanic trench. The Ordovician volcanic rocks in Maine continue from Newfoundland to central Massachusetts. Many geologists call it and associated rocks the Bronson Hill terrane. The Boundary Mountain terrane in Maine is part of the Bronson Hill terrane.

These volcanic rocks contain ore minerals, mainly copper, zinc, and lead. Mt. Chase contains an ore body that will probably support a mine someday.

Oakfield Hills

South of the interstate, near the Smyrna Mills exit, is the northernmost of a line of granite plutons that invaded the surrounding rocks during Devonian time. This line, which includes Katahdin, extends southwest to the New Hampshire border. The Oakfield Hills visible from the highway are a rim of baked slaty rocks that surround the Hunt Ridge and the Pleasant Lake granite plutons. The granites weathered and eroded into basins, while the more resistant baked slate around their margins stands as ridges. A similar, although not continuous, range of mountains southwest of Oakfield extends all the way to the White Mountains in New Hampshire. It includes all of the highest mountains in Maine except Katahdin.

Maine Potatoes—Rooted in Geology

Maine grows lots of potatoes. Those low bushy plants that cover the fields in Aroostook and Penobscot Counties are potatoes. Nowhere else in New England will you find anything comparable. Why?

The answer, of course, is rooted in the local geology. Some varieties of potatoes develop scab when the soil is too acid, or low pH, while other varieties develop scab if the soil is too alkaline, or high pH. Scab reduces the value of potatoes. In most New England soils the pH is so low that acid scab develops. Glacial activity mixed rock debris to form most Maine soils. The rocks in Aroostook County contain enough lime to make soils of about pH 6.5, neither too acid nor too alkaline.

Rocks rich in calcium carbonate, such as limestone and calcareous slate or sandstone, locally underlie large areas of north-central Maine. Weathering of the carbonate rocks produced soil in which potatoes thrive. In Aroostook County, along U.S. 1 between Caribou and Houlton, the potato fields are on the calcareous Carys Mills formation. Between East Corinth and Exeter on Maine 43 and 11, potatoes grow on soils formed on a limestone in the Waterville formation. In both areas outcrops are rarely seen; what lies beneath is expressed only by the vegetation. Exposures of calcareous rock can be identified by the distinctive tan color of their weathered surfaces, quite different from the rusty color of weathered sulfide rocks such as the Smalls Falls formation. Rocks that weather to greenish or silvery shades of gray are typically not calcareous.

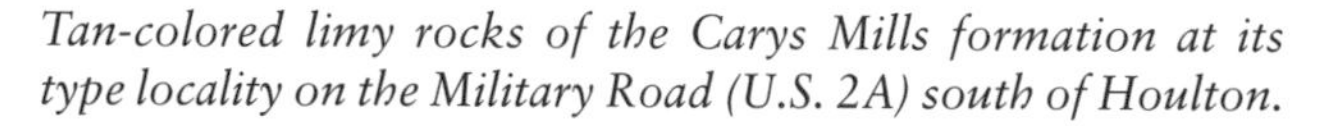

Tan-colored limy rocks of the Carys Mills formation at its type locality on the Military Road (U.S. 2A) south of Houlton.

Maine 7
Belfast—Dover-Foxcroft
62 miles

Maine 7 crosses rather monotonous rocks, most of which date from Silurian time, about 420 million years ago. Outcrops are few and widely spaced, with only an occasional high ridge giving some idea of what lies beneath the rolling lowlands landscape. Belfast is in the Avalon terrane; Dover-Foxcroft is in the heart of the Central Maine slate belt. Between them is the Norumbega fault zone, a line that represents a former ocean.

One advantage of this route is that it crosses the slate belt at right angles to the folds. That provides a greater variety in the rocks the road crosses. Too bad they are not well exposed.

Avalon Terrane

The rocks near the southern end of the route are part of the Avalon terrane, which consists mostly of rocks that were laid down as sediments during late Precambrian time. The road crosses rock units that include the rusty-weathering Penobscot formation, the calcareous mudstone of the Bucksport formation, and volcanic and sedimentary rocks of the Passagassawakeag formation. Small slivers of the Cape Elizabeth formation and the Bucksport formation appear near the Norumbega fault zone.

Norumbega Fault Zone

The contact of the Avalonian rocks with the Central Maine slate belt is a few miles north of Brooks. The Norumbega fault zone consists of three or four generally parallel faults; they are the weld that joined the Avalon terrane to Maine during Devonian time—during the Acadian mountain-building event. Earlier movements on those faults shoved the Avalon terrane over the edge of North America. Later movements slid it southeast, probably during the Alleghanian mountain-building event of Upper Carboniferous time.

Rocks near the Norumbega fault zone were crushed. Water penetrated along the fractures, weathering the rocks, making them more vulnerable to erosion. The streams that now follow them have developed a rectangular drainage pattern that alternates between following the faults and strong sets of fractures that intersect them at right angles. The rocks between the fault zones are quite durable and stand high as steep hills between the fault valleys.

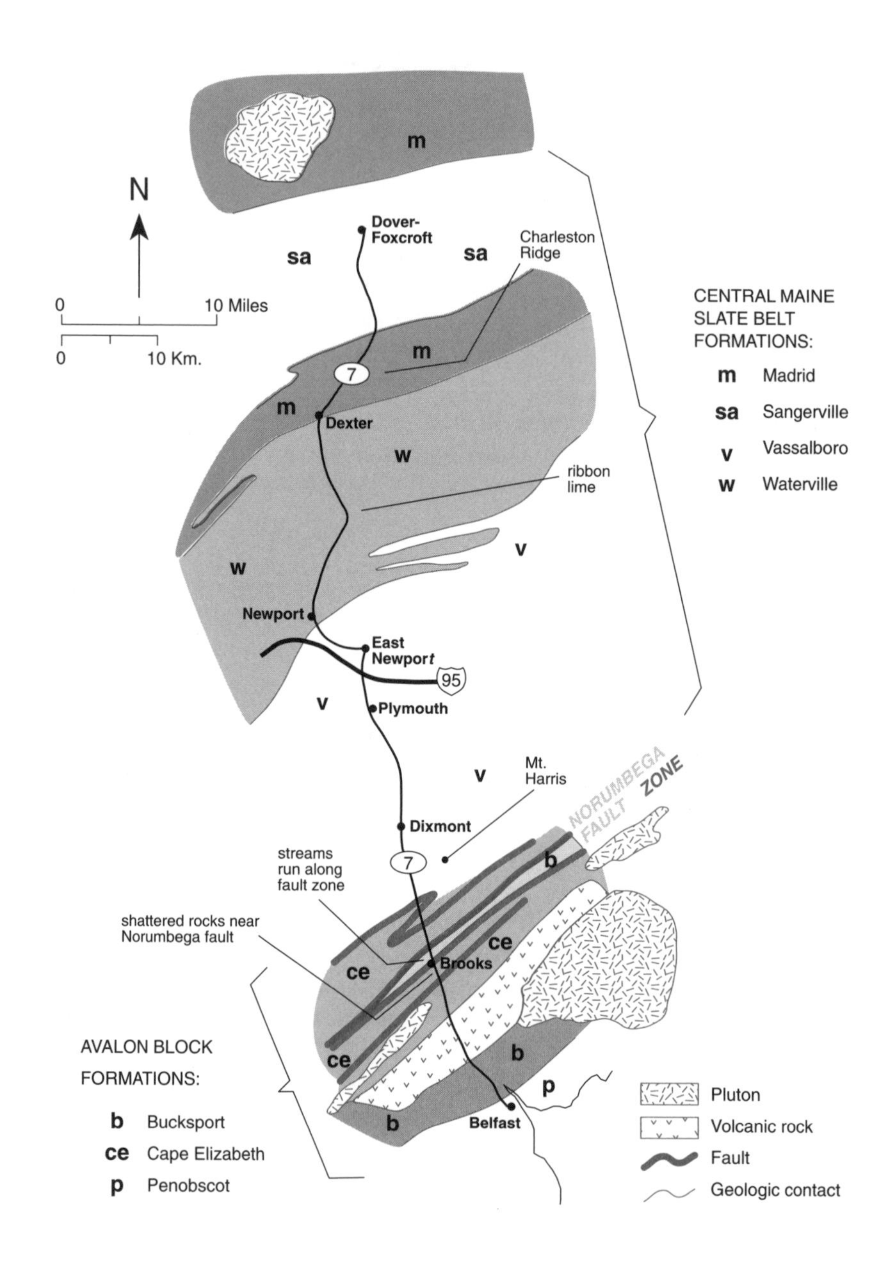

Geologic features along Maine 7 between Belfast and Dover-Foxcroft.

Deformed rocks of the Cape Elizabeth formation near the Norumbega fault in Jackson.

Vassalboro Formation

The Vassalboro formation lies immediately north of the Norumbega fault, along the southern edge of the Central Maine slate belt. It was deposited as muddy and limy sediments during Ordovician or Silurian time.

This Vassalboro formation includes metamorphosed limy mud and sandstone. Many outcrops expose alternating thin beds of these rocks. Sandstones consist mostly of quartz, which resists weathering and erosion. They support the summit of Mt. Harris in Dixmont.

Mt. Harris is a few miles north of the Norumbega fault zone. Except for Pleasant Mountain in Bridgeton, it is the highest elevation within the Central Maine slate belt. You can see a large part of Maine from the fire tower on its summit, including Mount Desert Island and Blue Hill to the southeast, the Camden Hills to the south, Mt. Waldo and Peaked Mountain to the east, Mt. Katahdin and the White Cap Range to the north, and, on especially clear days, Mt. Washington in New Hampshire. Between Mt. Harris and the mountains to the north is the broad lowland underlain by the easily eroded slates and limestone that make up the Central Maine slate belt.

Vassalboro formation exposed north of Mt. Harris in Dixmont.

Waterville Formation

At Newport, Maine 7 crosses from the Vassalboro formation to the Waterville formation, which is exposed between Newport and Dexter. The formation is named for Waterville, where it is exposed in outcrops of interbedded sandstone and slate. Parts of the formation consist of interbedded limy sandstone and slate, known in Maine as ribbon lime. The ribbon limes of the Waterville formation in central Maine and the Carys Mills formation in Aroostook County are so similar that they must have formed in the same manner.

Charleston Ridge

At Dexter, a nearly continuous ridge some 20 miles long breaks the monotony of the landscape of the Central Maine slate belt. This is the Charleston Ridge. It is composed of somewhat metamorphosed Madrid sandstone. Local tradition has it that an old silver mine is somewhere along its length.

A similar though less prominent ridge north of Dexter is also made of Madrid sandstone, on the other side of a synclinal trough. Think of it as a repetition of the Charleston Ridge.

Maine 23, the shortcut between Dexter and Guilford, crosses the second ridge of Madrid sandstone. At its crest, you can see a spectacular view of the central Maine highlands. Look for Mt. Katahdin and the peaks of the White Cap–Squaw Mountain Range.

Sangerville Formation

Bedrock between Dexter and Dover-Foxcroft is the Silurian Sangerville formation, a mudstone with more sand and silt than the shaly Waterville formation to the south. The Sangerville and Waterville formations also contain thin lenses of limy rocks, similar to those in the Vassalboro formation.

Maine 15
Bangor—Abbot Village
51 miles

Bedrock between Bangor and Kenduskeag is the Vassalboro formation, in places a calcareous mudstone known informally as ribbon lime. The limestone layers weather more rapidly than the sandy layers, and thus appear slightly indented on outcrops.

Bedrock north of East Corinth to Charleston is the Waterville formation, which contains thin but long lenses of ribbon lime. Between Charleston and Abbot Village, the bedrock consists of the Madrid and Sangerville formations. Both contain a good deal of quartz, which makes them more resistant to weathering than the calcareous rock to the south. Charleston Ridge is composed of the Madrid sandstone. To those driving north, it is a prominent topographic feature.

Seawater flooded all the area south of Charleston Ridge as the glacier of the last ice age melted. Sand, gravel, and clay deposited then now cover most of the bedrock. A small portion of the valley of the Piscataquis River north of Charleston Ridge was also flooded, probably no farther upstream than Dover-Foxcroft.

Kenduskeag Stream and Esker

Maine 15 follows Kenduskeag Stream as far as East Corinth, where it swings west toward Garland. During his second trip to Maine in 1854, Henry David Thoreau asked his Indian guides what the name of this stream meant in their language. Pulling his leg, they told him it

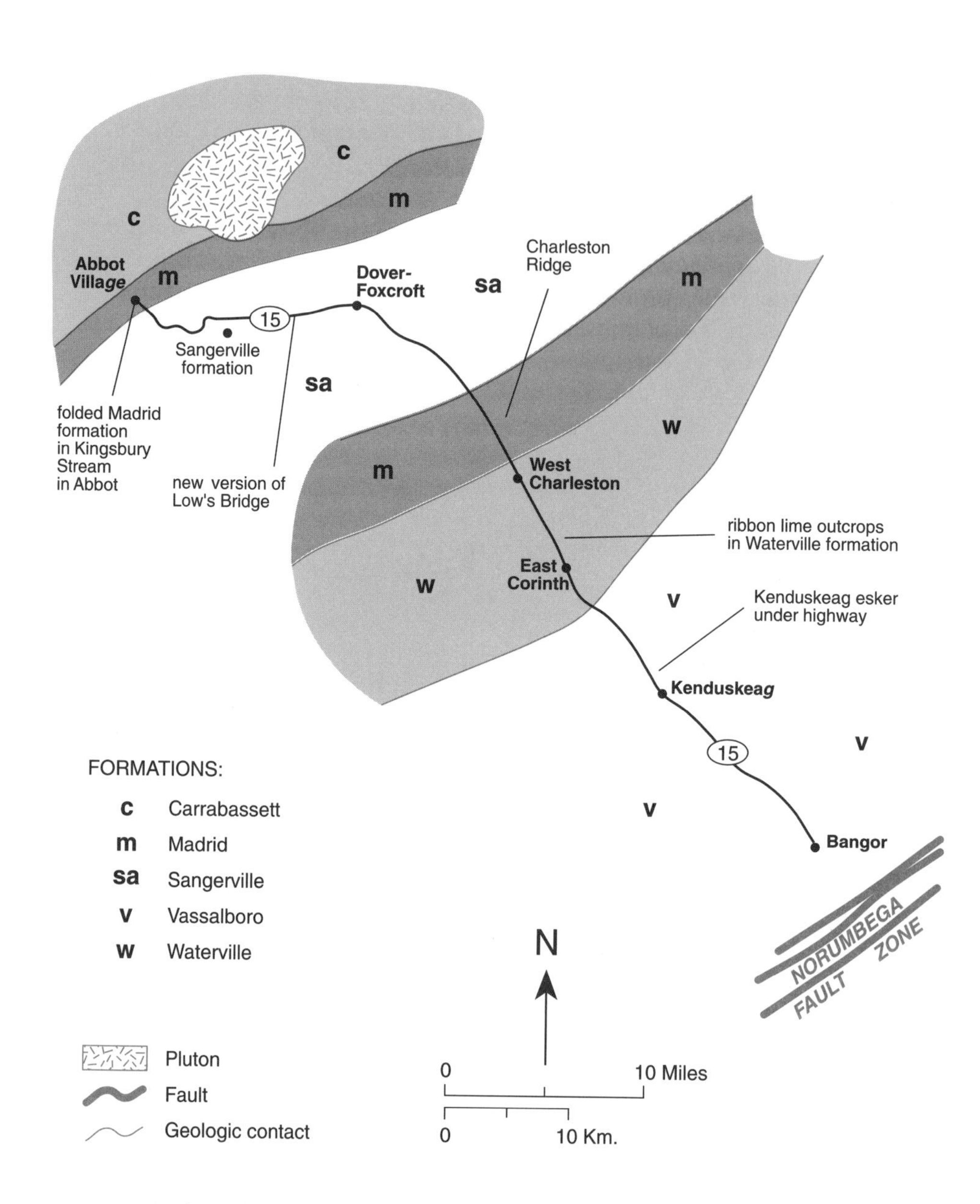

Geologic features along Maine 15 between Bangor and Abbot Village.

meant "when you paddle up the Penobscot past the mouth of this stream, you don't go that way." In his glossary of Indian words, Thoreau says it means "Little Eel River," so he was not taken in.

The road is built on top of an esker along much of the way between the villages of Kenduskeag and East Corinth. This esker is unlike most of those in Maine in that the ridge is generally less than 15 feet high and has a broad, rather flat top. These features likely mean that it was submerged soon after it formed, and waves eroded and flattened its crest. Whatever happened, its broad crest provides room for the highway and houses on either side. Most Maine eskers are so narrow that they barely have room for a road along the crest.

At East Corinth, the esker swings westward with a tributary of Kenduskeag Stream, where it passes through a notch in Charleston Ridge. At Sangerville, the esker swings into the valley of the Piscataquis River, which it then follows, with some breaks, to Moosehead Lake and beyond. South of Kenduskeag, the esker turns south toward Levant and Hermon, where it ends in a delta that the meltwater stream built into the sea.

Big House, Little House, Back House, Barn

Typical New England farmhouse construction in the 1800s consisted of several buildings connected to each other, and to the barn. Supposedly, this allowed the farmer to reach the barn without going out in the cold. That was an unplanned result. Commonly, a small house and separate barn were built. As the family grew, a larger house was built and smaller buildings were attached, until they finally connected to the barn. The final configuration typically consisted of four connected buildings, described in the ditty "Big house, little house, back house, barn." Watch for the fine example on the south side of East Corinth, west of Maine 15.

Waterville Formation and the East Corinth Farms

Weathering and glacial grinding of the ribbon lime of the Waterville formation produced some of the best agricultural soils in central Maine. In much of Maine, the last ice sheet ground up granite, sandstone, and shale to make loose deposits that form acid soils, generally poor for most crops except pine trees and strawberries. The Waterville formation added enough lime to raise the pH to about 6.5, making the soils suitable for growing several crops. The East Corinth region grows enough vegetables that several truck farmers are able to sell them in the less fortunate areas of Maine with poor soils. Watch between East Corinth and the Charleston Ridge for several large outcrops of ribbon lime in the Waterville formation.

Charleston Ridge

About 4 miles beyond East Corinth, the road climbs rather steeply to the crest of Charleston Ridge. The long ridge, with several other names between here and Dexter, is an outcrop of the Madrid formation, a sandstone deposited during Silurian time, about 420 million years ago. Watch for several outcrops of this rock along Maine 15 on both sides of the ridge.

This ridge commands an expansive view to the south toward the coast. The set of buildings near the summit was once a military radar installation and is now a prison. You can see a wide panorama of the mountains of Maine north of Charleston Ridge. These include the White Cap Range near Katahdin Iron Works, Borestone Mountain near Monson, and Big Squaw Mountain near Greenville.

Lime Kiln

Dover and Foxcroft were separate towns until they joined in 1922. A lime kiln preserved from the 1870s is a short distance from Center Road, near the Foxcroft Golf Course. The lime was mined from limestone lenses in the Silurian Sangerville formation and roasted to remove carbon dioxide, leaving calcium oxide. Agricultural lime has to be added to the generally acid soils of Maine. Early settlers produced their own until the railroads came through. Limestone was precious.

Ribbon lime of the Waterville formation in Charleston.

Low's Bridge

Low's Bridge, about midway between Dover and Guilford, is a rebuilt version of a long, covered truss bridge that was built in 1830. In 1987 a flood swept the bridge down the Piscataquis River. The replica installed in the summer of 1990 is probably the newest covered bridge in New England.

The flood occurred in April after about 8 inches of rain fell on a thick snowpack. Since no previous flood had removed Low's Bridge, the one of 1987 must have been at least a 150-year event. The U.S. Geological Survey estimated that it was the largest flood on the Piscataquis River in the past two hundred years.

Sangerville

South of the Maine 23 bridge over the Piscataquis River is the town of Sangerville, type locality of the mudstone of the same name. A low esker ridge, once the bed of a glacial meltwater stream, closely follows the south side of the Piscataquis River. Like many eskers, this one is now largely removed, sacrificed to the demand for sand and gravel.

Sandy mudstone beds of the Sangerville formation.

Sangerville was the birthplace of two men who became rich and famous. Hiram Maxim was an inventor who began tinkering in a small shop in Dexter when he was fourteen. His most famous invention was the machine gun, first tested at Abbot Hill in Dexter. He also held patents on smokeless powder, steam pumps, vacuum pumps, and an engine governor. Queen Victoria knighted him. Harry Oakes made his fortune in gold mining. King George V knighted him for his many contributions to charity. Sir Harry was ambassador to Bermuda when he met an untimely end at the hand of a jealous husband.

Madrid Formation at Abbot Village

Under the bridge over Kingsbury Stream is an interesting outcrop of the Madrid formation. The sandstone beds have tight and irregular folds with a few faults. As with most outcrops of the Madrid formation, they contain pods of such calcium silicate minerals as red garnet and green diopside. These, apparently, were some kind of calcium carbonate concretions that were metamorphosed along with their host rock.

Maine 11
Milo—Millinocket
45 miles

All bedrock along Maine 11 is either slate in the Carrabassett formation or sandstone in the Madrid formation. The Carrabassett slate was deposited as mud during late Devonian time; the Madrid sandstone was laid down during Silurian and Devonian time. The southern half of the road crosses deposits of sand and gravel—glacial outwash.

Marine Outwash Plain

Between Milo and Brownville, the road is mostly on a broad plain of sand and gravel. It was deposited at the end of the last ice age, when seawater flooded the valley of the Pleasant River for a time while the ice was melting. Meltwater streams dumped a large delta of glacial outwash in the flooded valley.

The low hills that rise above the delta surface near the river are segments of eskers, standing like islands above the flood of outwash

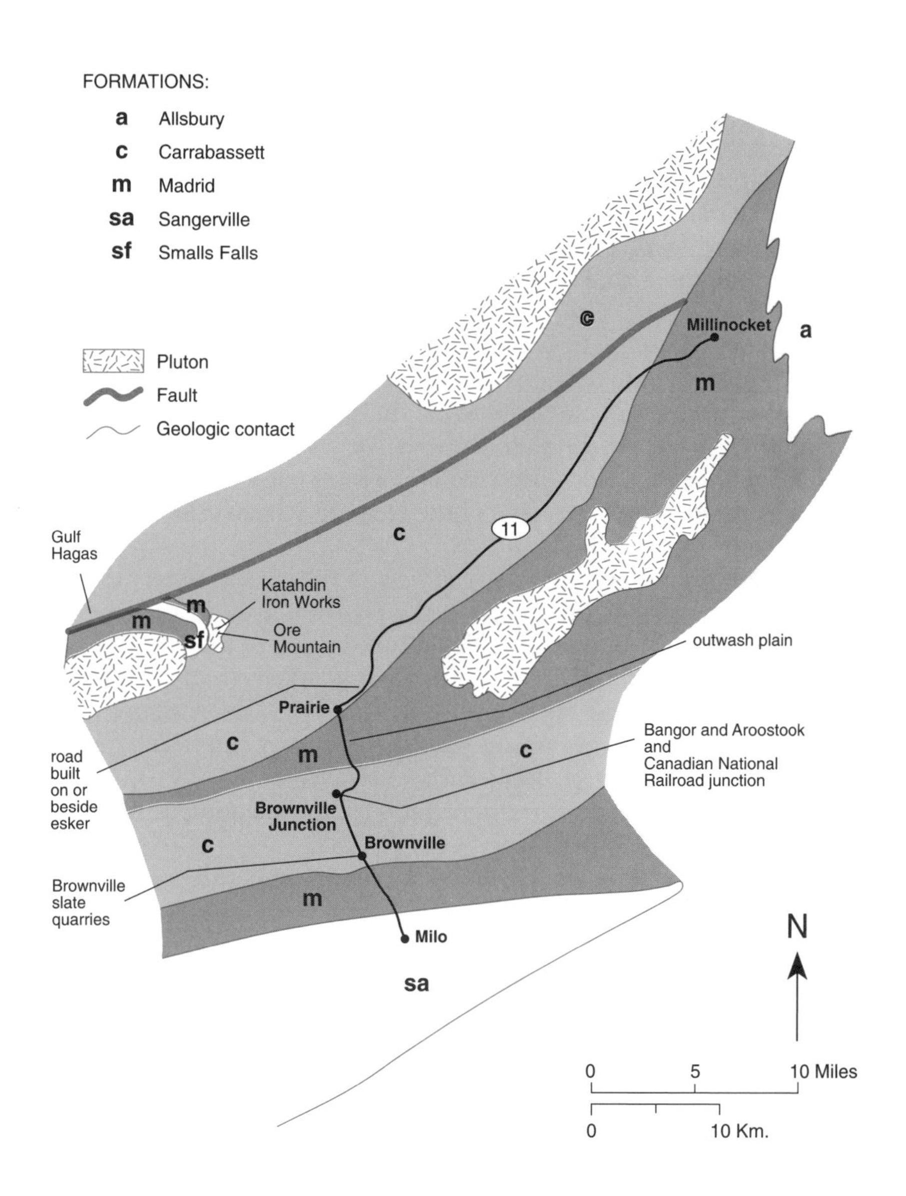

Bedrock geology between Milo and Millinocket.

in the delta. Glacial marine clay is believed to lie beneath much of this sand and gravel deposit.

Brownville Slate

A number of abandoned slate quarries in the Carrabassett formation near Brownville were part of a thriving industry, now defunct. Moses Greenleaf, an early entrepreneur interested in developing the resources of interior Maine, discovered high-quality slate in Brownville in 1814. Production began in the 1840s, when skilled quarrymen from Wales immigrated to Brownville and developed three quarries. One of the Welsh miners, Griffith Jones, discovered the more famous slate at Monson, whose quarries soon outstripped those at Brownville.

High-quality slate for pool tables and telephone switchboards was the major product of the Brownville quarries. Slate shingles were a by-product. Most of the slate shingles were shipped to Boston, but many roofs on the older Brownville homes wear these satiny black slates. A house on the east side of the Pleasant River is shingled completely with Brownville slate.

In 1984, bubbles of oil were seen rising through the watery depths of the Merrill quarry, the largest of the Brownville quarries, on a hill west of Maine 11. Bubbles, and local rumors, led to a salvage operation that involved the FBI, three million dollars' worth of equipment, and twenty-one divers. They recovered twenty cars, fewer than a third of them lying in a large conical pile on the quarry floor. The quarry was more than 300 feet deep, with water depths locally exceeding 150 feet. The cars, some not running too well, were dumped here and reported stolen to collect insurance.

Like the water in some basins in the ocean and large inland seas, such as the Black Sea, the water at the bottom of the quarry contained no oxygen. The cars that were recovered were remarkably well preserved, hardly rusted. Iron cannot oxidize without oxygen.

Prairie

Prairie is on the northern edge of Brownville, at the junction of Maine 11 and the Katahdin Iron Works Road. The name derives from the large fields here that are usually planted with wheat, oats, or potatoes, giving the land the look of a western prairie.

The flat surface is the top of a large delta. Meltwater streams built it as they deposited outwash sediments in the shallow seawater that flooded this part of Maine at the end of the last ice age. It marks the inland limit of the marine submergence in this area. The swampy ground just north of the road marks the location of the ice margin when the Prairie outwash was deposited. A few hundred yards to the northeast, Maine 11 is built on and beside an esker, the bed of the meltwater stream that carried the sand and gravel to the delta.

Katahdin Iron Works

The North Maine Woods gate to Katahdin Iron Works is about 6 miles along Katahdin Iron Works Road from the Prairie junction with Maine 11. The ironworks was the site of a mining operation that began in 1843 and continued sporadically until 1890. Plagued by high maintenance and transportation costs, fires, and an ever-fluctuating market, the ironworks was rarely without problems. When pig iron from the aging mill could not compete with modern steel production in Pennsylvania, the operation permanently closed. Only the large blast furnace and one of the sixteen charcoal kilns remain. Although the site had several other names, it is now known as the Katahdin Iron Works.

The modern discovery of the ore deposit, in the beginning of the nineteenth century, is attributed to Moses Greenleaf, who understood the original Abenaki Indian name for the mountain, *Munna Olammon Ungun,* meaning "area of much red pigment." The ore was a gossan, a thick soil composed largely of iron oxide. It capped Ore Mountain.

Gossans are extremely peculiar soils that develop on weathering rocks that contain large quantities of metal sulfide minerals. The Ore Mountain gossan developed on a large body of iron sulfide, composed mainly of the mineral pyrrhotite. Weathering oxidized the sulfide to sulfurous acid, which was carried away in solution. Meanwhile the iron became iron oxide, which remained in the soil as the minerals hematite and limonite. The hill acquired a carapace of red gossan soil. Indians traveling up the Pleasant River must have noticed the red gossan on the north slope of Ore Mountain. And they may have used it.

Buried deep in the annals of Maine history and folklore are stories of the Red Paint People, a highly sophisticated Indian maritime community that buried its dead in graves lined with red ocher—hematite pigment. Red ocher graves four thousand to seventy-five hundred years old have been found in coastal regions from Labrador and Newfoundland to southern New England. Bones dissolve readily in the acid soils of Maine, so red ocher and stone tools are commonly all that remain. Only where graves lay beneath shell heaps was the soil neutralized enough for bones to survive.

Today, a visit to the top of Ore Mountain is like visiting a Martian landscape, at once desolate and beautiful. One can learn a lot by studying the slopes of this mountain. Where removal of the gossan exposed the pyrrhotite, it is now weathering, producing extremely acid water that poisons the soil and streams. In some areas, the soil is so acid that only scrub birch grow. As the acid water flows down the slope, it dissolves more minerals and saturates dead leaves, wood, and other organic matter. Reactions between the organic carbon compounds causes the iron to precipitate in the leaves, cones, and sticks. Leaves, cones, and wood become heavy and brittle as they start down the road to fossilization. The precipitated iron is called ferricrete. Drainage from Ore Mountain has stained the boulders in the Pleasant River reddish brown.

The pyrrhotite body beneath Ore Mountain reaches a depth of more than 1,000 feet; it may well be the largest massive sulfide deposit in the East. The pyrrhotite formed as the molten diorite magma of the Moxie pluton invaded the Smalls Falls formation during Devonian time. The Smalls Falls formation is dark with organic matter and finely disseminated iron oxide. As it began to melt in the extreme heat of the diorite magma, it formed a molten sulfide magma that separated from the basalt magma in the same way that oil separates from water. The sulfide magma crystallized into a massive body of pyrrhotite.

Scene near the top of Ore Mountain, Katahdin Iron Works. The dark rocks are massive sulfide, coated with dark brown iron oxide. The soil is a remnant of the red gossan. —Marvin Schmid photo

Unlike many other pyrrhotite bodies, the one at Ore Mountain contains no nickel, nor is it magnetic. It does contain cobalt and other rare metals, perhaps in minable concentrations. Mining companies have drilled a number of exploratory holes.

Gorgeous Gulf Hagas

Gulf Hagas Gorge is upstream from Katahdin Iron Works, along the West Branch Pleasant River. It is certainly the longest gorge in Maine, and arguably the most spectacular. Water flows between vertical walls of slate more than 120 feet high for more than 3 miles and tumbles down waterfalls as much as 25 feet high. The total drop over the length of the gorge is about 370 feet. The lower walls and floor of the gorge contain large potholes that were eroded as persistent eddies turned pebbles on the bedrock floor of the stream.

Early log drivers created jagged exposures of bedrock deep in the gorge, where they blasted obstructions to the flow of logs. The Appalachian Trail follows the length of the gorge on its north side. Hay Brook Falls, on a tributary downstream from Gulf Hagas, is a small gorge with beautifully clear water and potholes. It cuts through a chaotic unit of the Carrabassett formation.

Not far from Gulf Hagas, on the north side of the West Branch Pleasant River, is the Hermitage, where a set of sporting camps flourished near the turn of the century. The Hermitage is now the site of one of the few remaining stands of virgin pine in New England. Giant pine and hemlock, up to 10 feet in diameter, tower 150 feet or more over a soft and uncluttered carpet of pine needles. It is one of the few remnants of the magnificent forest that covered New England just a few centuries ago.

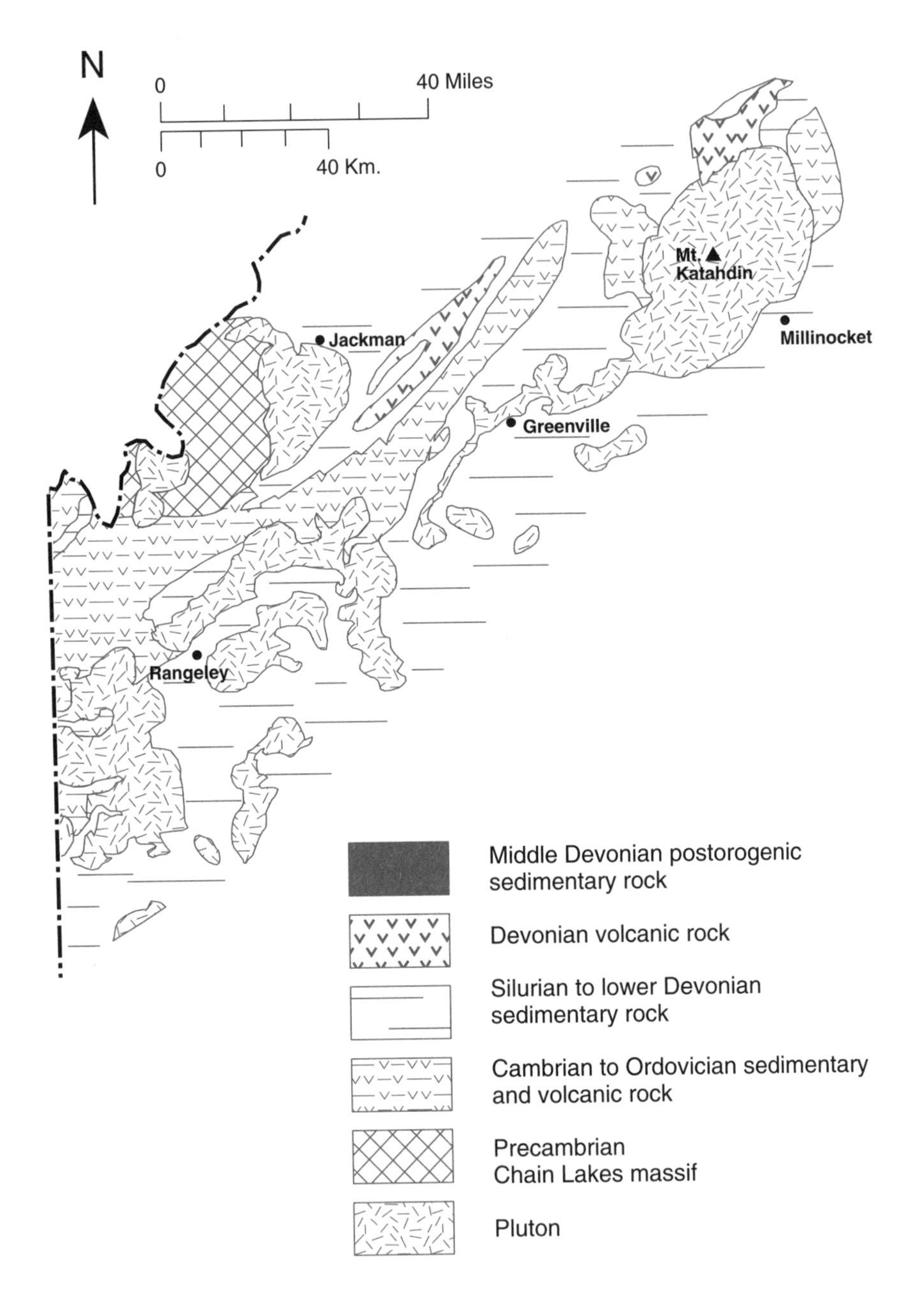

Geology of the mountains of Maine.

The Mountains of Maine

The highest mountains in Maine are in a belt that extends from the New Hampshire border northeast beyond Katahdin, more than 150 miles. While Katahdin is its highest point, the range generally becomes higher toward the southwest. It continues through the White Mountains of New Hampshire, then to Mt. Monadnock in the southwestern part of the state. The total length of the range is almost 300 miles.

Most of the rocks in these highlands are Devonian in age. The Acadian plate collision raised them into mountains. Volcanic rocks in this mountain belt erupted during the Taconic mountain-building event, then were deformed in the Acadian event. The mountains of today consist of rocks that were deeply buried during Devonian time, to depths of about 3 miles near Katahdin and to more than 10 miles in western Maine, where the Mahoosuc Range crosses into New Hampshire.

In other sections of this book I subdivide Maine into regions that differ in their bedrock. Only their landscape separates the mountains from the rest of Maine. The central Maine lowlands abruptly terminate against this mountainous belt, but the rocks are similar in both areas. The Carrabassett formation, for example, is widespread in both the lowlands and the highlands. In almost every way the formation in both areas is identical, except that in the mountains it was baked in the heat of magmatic intrusions, during Devonian time.

Ordovician Volcanic Rocks

Although rocks of Devonian age underlie most of the highlands of Maine, the region also contains volcanic rocks that erupted during the Taconic collision of Ordovician time. These are extensions of the Bronson Hill terrane of southern New England.

These older rocks are exposed in the cores of anticlinal arches, including the Boundary Mountain fold belt and the Lobster Mountain anticline. The arching fold of an anticline raises older rocks to a level with younger rocks on either side. As erosion planes the fold, it exposes the older rocks in the core of the fold.

The Ordovician rocks have been folded at least twice, so the slates have two or more planes of cleavage. They tend to break into blocks instead of simple sheets. Such heavily fractured slates are relatively weak; they generally erode into low areas that contain river valleys and lake basins.

Strange Massif at Chain Lakes

The Chain Lakes massif is a mountainous area near the Québec border that consists of ancient gneissic rocks. Recent work suggests the Chain Lakes massif is an exposure of Grenville gneiss. Previous dating of radioactive minerals in the gneiss indicated an age of 1.6 billion years, about half a billion years older than the Grenville gneiss. Grenville rocks are exposed in Québec north of the St. Lawrence River and in large areas of the Appalachians, but not in Maine. They may underlie the Chain Lakes massif itself and the rest of the rocks in northern Maine.

No one seems to have a good explanation for the origin of the Chain Lakes rocks. This is partly because they are so thoroughly metamorphosed that it is almost impossible to know what the original rocks were. Some geologists believe the Chain Lakes massif may be the site of a large asteroid impact. The scant evidence for this includes layering and a variety of clasts set in finer rock, thought to form from the fallback of erupted material. Some microscopic features in these rocks and shatter cones are interpreted as shock structures from the impact. The lower part of the Chain Lake massif is described as the target rock, while the middle layers are thought to have formed during the accumulation of debris thrown up by the impact. The youngest portion of this massif consists of sedimentary and volcanic rocks accumulated after the impact event.

Boil Mountain Ophiolite and Penobscot Mountain-Building Event

Ophiolites are slices of ocean floor that were shoved onto a continental margin. An ophiolite typically consists of several distinct layers: an upper layer of sediments; below them, pillow basalts erupted at an oceanic rift; below that, the dikes that fed the basalt flows; and a lower layer of gabbro crystallized from basalt magma pooled below the ocean floor.

The Boil Mountain ophiolite was probably on the floor of the ocean during Cambrian time, to judge from fossils in an overlying formation. It was shoved onto the Chain Lakes massif during early Ordovician time. This collision of oceanic and continental crust is called the Penobscot mountain-building event. The force of this collision likely buckled the Grenville basement, making it possible for the seafloor to override it.

Another, much larger belt of ophiolites lies about 50 miles to the north in Québec. Those were shoved onto North America during late Ordovician time, during the collision between ocean floor and continent that geologists call the Taconic mountain-building event. The Taconic event may have moved the Chain Lakes massif and its overlying ophiolite close to its present location.

Highlands and Lowlands

Devonian slates, granites, gabbros, and volcanic rocks are common in the highlands of Maine. The volcanic rocks stand high because they resist weathering and erosion; the plutonic rocks, granite and gabbro, are not so resistant and tend to erode into lowlands. Katahdin is an exception; it is an unusual kind of granite that does resist weathering and erosion. The other high peaks are eroded in baked slates of the Carrabassett formation.

Mudstone, Slate, and Hornfels

When mudstone is folded while it is buried deeply enough to reach temperatures of at least several hundred degrees, it turns into slate. The distinctive feature of slate is its tendency to break along parallel surfaces to make thin sheets. It does that because the mineral grains that crystallized while it was being folded align parallel to each other, in military formation, as it were.

If mudstone or slate is simply heated, it bakes into a rock called hornfels, in which the mineral grains are not in parallel alignment. This typically happens around the margins of an igneous intrusion, where the enclosing rocks bake in the heat of the magma. The mineral grains that grow while hornfels are baking tend to make spots—typically pink andalusite and red cordierite grains—in the rock. Most of the rocks have a sparkled purplish color from biotite flakes. Geologists tend to utter the phrase "spotted hornfels" as though it were one word. Hornfels do not split into slabs and sheets in the manner of slate.

Slates tend to weather fairly readily because water seeps into them along the fracture surfaces. Hornfels have fewer fracture surfaces, so they tend to resist weathering, even though their chemical composition is the same as that of slate. In the lowlands of central Maine and in the transition to the highlands, the Carrabassett formation is a slate.

In the mountains, the Carrabassett formation is a hornfels. That, more than any other factor, explains the difference between the lowlands and the highlands.

Glacial Features in the Mountains of Maine

Some of the higher mountains caught enough snow during the ice ages to support small valley glaciers. Those include Katahdin, Sugarloaf, and Old Speck, all of which have cirques on them. A cirque is a deep bowl gouged in the head of a glaciated valley. Glaciers carve them as they freeze tight to the bedrock, then pluck blocks out as they move down the slope.

Arctic Plants and Patterned Ground

Plants typical of the arctic tundra grow above tree line on the high mountains of Maine. They are probably remnants of a tundra in front of the great continental glacier of the last ice age. When that ice in Maine finally melted about 12,000 years ago, tundra plants briefly covered all of the state. As the climate warmed in the lowlands, tundra vegetation survived only on high mountain peaks where the climate suited them. The mountaintop climate is too severe for the plants—the pine and hardwoods particularly—that now thrive in lower regions. The present climate on these peaks is similar to that of the arctic, with average temperatures at or below the freezing point of water. A warmer climate would allow lowland plants to colonize the uplands, and the arctic flora could not compete.

Drilling operations revealed that two high peaks in New England with arctic plants, Sugarloaf Mountain in Maine and Mt. Washington in New Hampshire, have permanently frozen ground. Permanently frozen ground tends to become patterned, to develop arrangements of stones into rings, stripes, or festoons. The existence of patterned ground on many of the high peaks of Maine suggests that they also have permanently frozen ground. But it is also possible that the patterned ground is a relic of the ice ages, not an aspect of the modern scene.

A Brief History of Logging in the Maine Woods

Only twenty years after the Pilgrims arrived in 1620, the first sawmills were in operation in York and Berwick. From then until after the Civil War, pine was the major concern of the Maine logger. Pine was used for lumber, and tall, straight pine was used for masts for the king's navy. Resentment over the king's pine was but one of the issues that led the thirteen colonies to the Revolution, but in Maine it may have been the major issue. The Aroostook War, between 1839 and 1842, was a dispute over timber rights.

Portland was the center of the lumber trade in the early eighteenth century, and the trade spread slowly northeastward along the major rivers to the Androscoggin, the Kennebec, and finally the Penobscot. Bangor and the Penobscot River became the capital of the lumber drive during the nineteenth century, moving up to the headwaters region in the East and West Branch by the 1830s. The West Branch River Driving Company sent long pine down the river, which was damned in places so that water could be released to wash down the logs. Loggers lived in the woods from summer to late winter and then took the logs with them with the spring snowmelt. They raised hell in Bangor till their money was gone and then were dragged back to the woods by the crew bosses. Loggers who came to Bangor, from the West Branch or elsewhere, were known as Bangor Tigers.

The woods camps were provisioned by wagons and sleds pulled by oxen and horse along tote roads. There were a number of "farms" in the woods, large clearings where fodder was grown for the horses and oxen that moved the logs to water. One of these, Pittston Farm on Seboomook Lake, until recently operated as a Boy Scout camp and is now a popular inn. Remnants of others also can be found, including Grant Farm about halfway between Greenville and Ripogenus Dam and Trout Brook Farm in the northern part of Baxter State Park.

Paper replaced lumber as the major wood product by the last quarter of the nineteenth century. The Great Northern Paper Company bought out the West Branch River Driving Company and built some of the largest paper mills in the country in Millinocket and East Millinocket. These towns were created in the middle of the northern woods, resembling the isolated mining towns in the West. Other paper mills are farther south on the major rivers: Boise Cascade and International Paper on the Androscoggin, Scott Paper on the Kennebec, and Champion International at the mouth of the Penobscot at Bucksport. The pulp wood for these mills comes from the north woods, from both the mountains and farther north in the wilderness of Maine. Georgia-Pacific has a mill on the St. Croix River, with the pulp coming from the unorganized townships of eastern Maine.

By the early 1900s, a few all-weather roads began to penetrate the wilderness and to replace the old tote roads. One was built from Moosehead Lake to Ripogenus Dam to bring in material for that dam, and later a power boat, the *Tethys,* was hauled over the same road.

West Branch No. 2 *towing on Ambajejus Lake in the 1950s.*
—Great Northern Paper Company photo

This towboat was used for hauling booms of pulp logs down the 20-mile-long Chesuncook Lake. A boom is a ring of logs, generally one hundred or so of them, chained end-to-end and holding four to five thousand cords. Under tow a single boom was liable to break, either at the chains or the drilled hole the boom chain pulled through, so double booms, one outside the other, were commonly used for large loads.

In the rivers the logs went free with the current, usually augmented with water released from dams. This extra flow was known as a driving pitch. As the logs came into the lake downstream, they were captured by booms extended across the mouth of the river. Booms were then made up and towed down the lake to the next dam. The West Branch and other major logging rivers carried wood belonging to several companies, so that all logs were branded, like cattle. When a drive reached a mill, say at Millinocket, the Great Northern logs were sorted out and the rest of the logs continued down the river. All log drives ended on the West Branch and other rivers in Maine in the 1970s and all logs are now trucked to the mills.

U.S. 2
Skowhegan—Bethel
80 miles

This section of U.S. 2 lies close to the border of the Central Maine slate belt to the south and the mountains of Maine to the north. Between about Wilton and Bethel, this road is well within the mountains. The mountains are readily visible from most of the rest of this section of U.S. 2. For these reasons, I have decided to place the description of the road here in order not to divide it into two sections of the book.

This portion of U.S. 2 traverses folded sedimentary rocks more or less parallel with the northeastward trend of the folds. They were deposited during Silurian and Devonian time and folded during the Acadian collision in the Devonian period. Near Skowhegan, they are now slates and phyllites, only slightly recrystallized.

The degree of metamorphism increases westward, where similar rocks were more deeply buried and absorbed some heat from the Sebago granite pluton. Near Bethel the rock formations that are slates near Skowhegan have been metamorphosed to schists that contain the fibrous mineral sillimanite, which forms only at temperatures that approach the melting point of the rock.

Between Skowhegan and Farmington, the route crosses glacial marine deposits laid down when the coastal region of Maine was submerged at the end of the last ice age. The rest of the route is above the limit of marine submergence. Small valley glaciers and the regional ice sheets both eroded the high mountains of western Maine, contributing to their landscape.

A Rusty Tillite

Watch between Skowhegan and Norridgewock for a rusty outcrop of glacial till, a relic of the last ice sheet. It contains boulders of the Smalls Falls formation, which contain pyrrhotite, an iron sulfide mineral. On exposure to air and water, the iron sulfide oxidizes to rusty iron oxide. In this case, it is so abundant that it cemented the till into a tough rock. You need a hammer to break it.

The Death of Father Rasle

One of the nastiest events in the history of Maine happened in Norridgewock. A Jesuit priest, Father Sebastian Rasle, who was stationed here to minister to the Abenaki Indians, was murdered by a

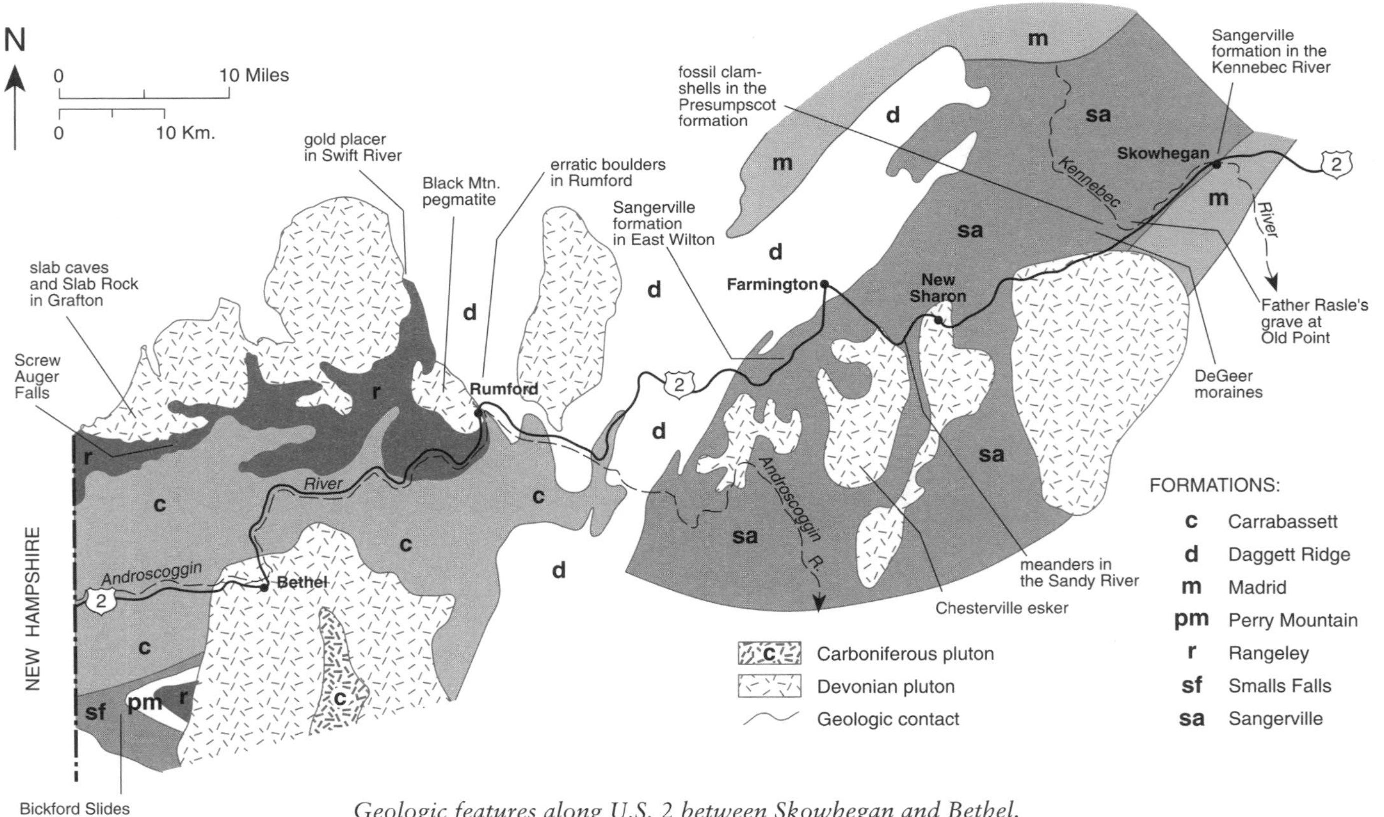

Geologic features along U.S. 2 between Skowhegan and Bethel.

Slate and sandstone of the Sangerville formation exposed in Kennebec River at Skowhegan.

band of British and Americans in 1724. Father Rasle had arrived in 1696 and was compiling a dictionary of the Abenaki language while maintaining his small parish. During the raid in which he was killed, the village and the dictionary were destroyed and the Indians slaughtered.

Father Rasle's grave is where he died, on Old Point near the confluence of the Sandy River with the Kennebec. Benedict Arnold visited the grave on his way to Québec City in 1775 to attack the British. When last heard from, Father Rasle was a candidate for sainthood.

Kennebec Esker Buried in the Mud

Just northwest of Norridgewock, a large esker leaves the Kennebec and continues southward toward Belgrade and its end at the Augusta airport delta. The esker is draped with the mud of the Presumpscot formation. The esker was first formed in the usual manner in a tunnel beneath the last ice sheet. Then, as the ice retreated, waters of the marine submergence covered the esker, partly burying it with the mud of the Presumpscot formation. The Presumpscot silt and clay here

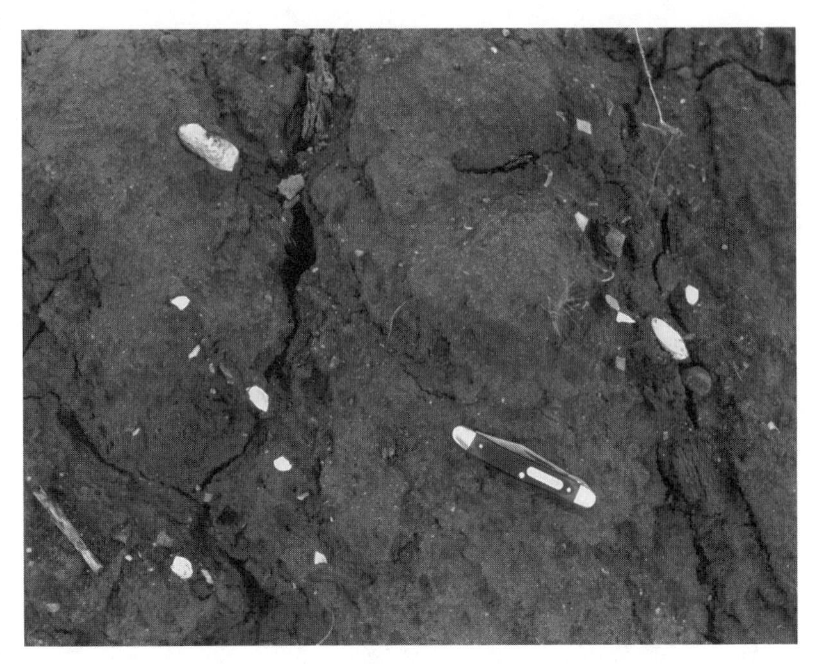

Clamshells in the Presumpscot formation, Norridgewock. These shells are about 12,500 years old.

contains abundant clam and snail shells. Similar shells from this site were some of the first items in Maine dated by the then-new carbon-14 method. The Norridgewock shells were dated at about 11,500 years old. It is now understood that outer portions of shells like these were contaminated with younger, more radioactive carbon-14. The actual age of the shells is about one thousand to fifteen hundred years older than the original dating suggested.

A huge pit was excavated in this esker during World War II, the sand and gravel being used to construct an emergency landing field nearby. This field, which was designed for aircraft returning from Europe, is said to have cost one million dollars. Only one plane ever used it for an emergency landing during the war, but that one plane had a replacement value of over three million dollars—making the landing strip well worth its cost. The strip was sold to Norridgewock for one dollar and is used for private planes and an occasional car race.

DeGeer Moraines

Within a mile to the west of Norridgewock village, U.S. 2 passes fields with numerous rounded, short ridges—DeGeer moraines. They mark places where the last ice sheet grounded on the ocean floor in about 200 feet of water. The ridges record successive positions of the melting ice margin, between about 13,000 and 12,500 years ago.

Belgrade Lakes and the Rome Pluton

Four of the Belgrade Lakes are in basins eroded in the Rome granite, which is Devonian in age, about 380 million years old. As in many granites in Maine, the mineral grains in the Rome granite do not interlock. The grains of feldspar and quartz merely lie next to each other, like a jigsaw puzzle with square and rectangular pieces. When such a rock weathers, the mineral grains simply fall apart, forming a sandy soil. Glacial erosion probably made the actual basins, but the rock was weakened by millions of years of weathering before the ice got here. On the other hand, small stringers of Rome granite support portions of the highest elevations in this area, on Vienna and McGaffey Mountains.

The boulders in the photograph on page 13 are Rome granite, but since the underlying bedrock consists of the same rock, these are not true glacial erratics.

DeGeer moraines along U.S. 2, west of Norridgewock.

The Belgrade Lakes have long been a popular destination for family summer vacations and support a number of summer camps for children and teenagers. As a child, the great American writer E. B. White summered here with his family. A salon for fashionable ladies operated on Long Pond until recently.

New Sharon Glacial Deposits

Downstream from the U.S. 2 bridge over the Sandy River is a high bank faced with granite blocks. The bank has been collapsing as the river eroded its base, threatening the section of Maine 134 that runs along its edge. The granite block riprap was installed in an effort to stop the erosion. A student of mine after a field trip here reported: "They piled up riffraff in New Sharon to stop the erosion." That just might work, too.

The high bank at New Sharon is a complex deposit of glacial till and thin-bedded clay layers. An ancient lake or swamp deposit near its base contains numerous spruce logs more than 50,000 years old. A glacier, perhaps the last one through here, either buried this swampy material or eroded it from somewhere else and left it here.

The exposure of this section of glacial deposits apparently began during the great flood of March 1936, one of the largest on record in many New England streams. A mile or so downstream from the bridge, the Sandy River looped around a high, thin bank of glacial deposits. During the flood, the river made a shortcut right through this bank, collapsing it in the middle and leaving part on the other side of the river. The shorter path the river took here gave it a steeper gradient, causing the stream to erode its channel, exposing the unusual glacial deposits along its banks.

Chesterville Esker and the Sandy River Sand Dunes

For about 6 miles, a paved road south of Farmington Falls follows the crest of the Chesterville esker, also locally called the Devil's Backbone. It provides a dry and scenic passage through an area of many ponds and swamps. In places, the narrow ridge is at least 100 feet above the wet lowland. The ridge becomes lower and discontinuous toward the south. In Fayette, near Twelve Corners—where three roads cross—the Chesterville esker is covered by glacial marine mud. It continues as a much more subdued ridge for another 15 miles and finally ends in a glacial marine delta near Leeds.

The area near Chesterville is locally known as Egypt because it contains so much sand. Sand is also plentiful in the Sandy River valley to the north, between Farmington and New Sharon. After the ice finally melted off this area, about 12,500 years ago, it was barren until plants covered it a few hundred years later. During this time the wind blew the glacial sand into dunes. The strong northwest winds blew some of the dunes out of the valley, to Cape Cod Hill, which has a dune about 3 miles long on its western slope. Plants eventually stabilized the sand.

Settlers cleared the forest around 1700, plowed the land, and introduced grazing livestock. Large areas of old dunes were again exposed to the wind, which again drove the sand before it. After the Civil War, many people left the farms in New England, which promptly grew a second growth of trees. They again stabilized the sand. Holes dug into the sand commonly show a modern soil at the surface weakly developed on the sand that blew before the Civil War. Below that sand is the much better soil that developed between about 12,000 years ago and A.D. 1700.

Maine still has many areas of active sand dunes: in Wayne, in an area near Solon on the Kennebec River, and in the Desert of Maine in Freeport. All of these are glacial outwash deposits that lost their soil cover, baring the sand to the wind.

Floodplain of the Sandy River

Upstream of the New Sharon gorge, the Sandy River meanders about on a broad floodplain. The main flow of the stream tends to hug the outsides of the meander bends, called the cutbank. The river cuts out sediment here during floods. Meanwhile, sediment accumulates on the insides of the bends, called the point bar. Thus, the stream constantly tends to migrate toward the cutbanks on the outsides of the bends. As that continues, the bends become more pronounced. Eventually, the stream takes a shortcut across the neck of the bend, leaving it isolated on the floodplain. The section of the Sandy River between New Sharon and Farmington has gorgeous meanders and abandoned meanders.

Farther upstream, beyond Farmington, the valley is so narrow that the meanders extend from one side of the floodplain to the other, in many places cutting into the valley walls. The isolated pieces of floodplain are known locally as intervales.

Sangerville Formation

Between Skowhegan and Wilton, most of the rocks along U.S. 2 belong to the Sangerville formation of Silurian age. Near East Wilton it consists of thick-bedded sandstones that don't show the effects of

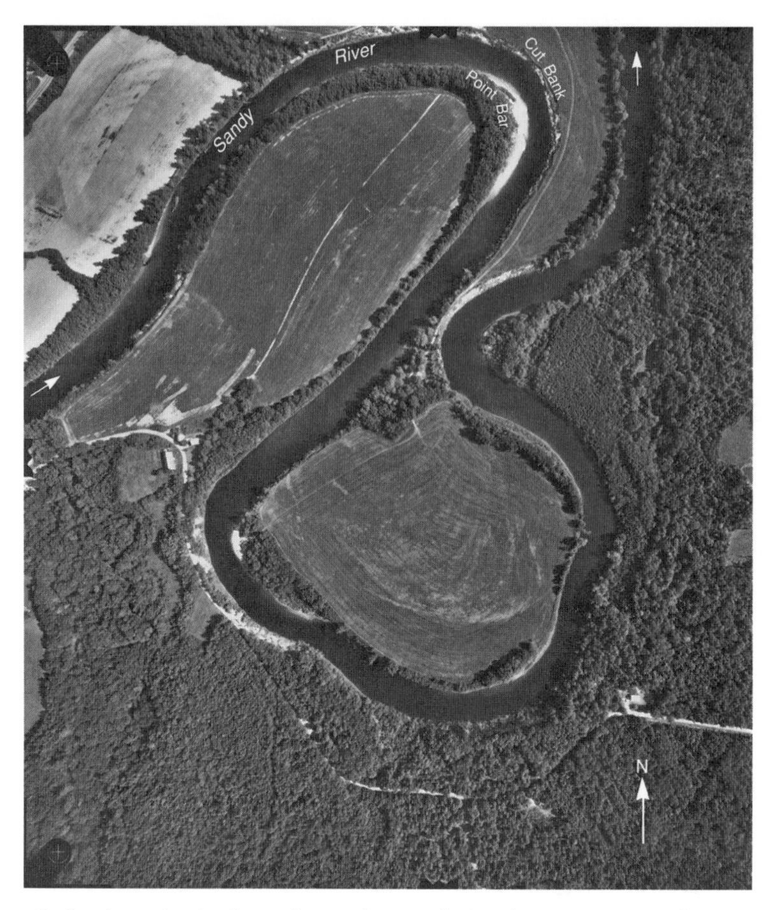

Cutbanks and point bars of meanders on the Sandy River in New Sharon.
—Sewall Company photo

metamorphism as much as slaty rocks do. Just north of here the Devonian Carrabassett formation is the most common bedrock.

Jay Granite Quarries

You can see the old Jay quarries on a high ridge near North Jay. They shipped granite by rail along the East Coast and supplied building foundations for many older homes. Many of the quarrymen who worked here emigrated from Italy.

Bald Mountain

The bare summit of Bald Mountain is an expanse of glacially polished and striated pavement eroded on the Carrabassett formation. It provides a marvelous view of the folds in the slate, which differ from most of those you see in this formation.

Sandstones of the Sangerville formation in East Wilton.

Most of the small folds in the Carrabassett formation appear to have formed when the sediments were still soft. They slid down submarine slopes. In many outcrops you can see tightly folded beds sandwiched between others that are not folded. And the slaty cleavage does not follow a direction parallel to a plane that would slice the folds into mirror images.

The complicated folds exposed in the glaciated pavement on Bald Mountain appear to have formed as the rock was compressed during the Acadian mountain-building event. All of the layers of sediment are similarly folded. And you could, at least in principle, slice any one of the folds into symmetrical halves by breaking it along the right plane of the slaty cleavage.

Mt. Blue

The distinctive steep south slope of Mt. Blue is easy to recognize from many places to the south and east. Glacial erosion of contrasting kinds of bedrock gave the mountain its shape.

Glaciers flowing south over the top of Mt. Blue froze tight to the rock on its south side, then plucked out blocks that broke along the pattern of fractures. That gave the mountain its steep southern slope. And that slope has a ragged look, the usual result of glacial plucking. Meanwhile, rocks embedded in the flowing ice abraded the northern slope, sandpapering it to a much smoother and gentler profile.

The bedrock that holds up the highest elevations on Mt. Blue is baked hornfels in the Carrabassett formation. Like those on Bald Mountain, these rocks were tightly folded during the Acadian mountain-building event. Hornfels in the Carrabassett formation, baked in the heat of nearby igneous intrusions, is the bedrock in most of the high mountains to the northeast.

While the hornfels resist erosion to stand high as peaks, the igneous intrusions that baked them weather and erode into lowlands. The Phillips pluton, a granodiorite emplaced during Devonian time, is a good example. It lies beneath Lake Webb and its surrounding valley. The quartz grains in granite become sand grains as it weathers into soil. They found their way into large areas of sandy glacial outwash, and from there into the sandy beaches around Lake Webb.

Tumbledown Mountain

Maine has two Tumbledown Mountains, and three Tumbledown Dick Mountains. Tumbledown Dick apparently means fallen rock. This Tumbledown is near Mt. Blue, along the road between Weld Corner and Byron on the Swift River, in Township 6 North of Weld. Bedrock in this Tumbledown is much like that in Mt. Blue, turbidite mudstones of the Carrabassett formation baked to a hornfels in the heat of the Phillips pluton.

Glacial ice rode over the mountain, giving it a steep south face and a gentle north slope. It has two peaks with a beautiful alpine lake between them. Maine has a few lakes higher than Tumbledown Pond, but none are more beautiful. The steep southwest side of the mountain is suitable only for rock climbers. The Brook Trail and the Parker Ridge Trail both lead to Tumbledown Pond, with a climb of about 2,000 feet in about a mile and a half.

Swift River Gold Placer

The legend of the discovery of gold in the Swift River reads like a headline on a supermarket tabloid: MAINE CHICKEN DISCOVERS GOLD: OWNER BECOMES WEALTHY WHILE CHICKEN BECOMES DINNER.

Tradition has it that a farmer used gravel from the Swift River to provide grit for his chickens. While preparing one for Sunday dinner, he found a gold nugget in its gizzard. I think it more likely that someone home from the California gold rush tried his pan in the stream and discovered the Swift River gold in the late 1850s. A similar tradition has chickens finding diamonds in the glacial deposits of Wisconsin.

The Swift River gold deposit is a placer, a concentration of heavy minerals in a streambed. Placer gold is eroded from some other source, usually a bedrock quartz vein called the mother lode, then concentrated in the stream channel. So far, no one has found the source of the gold in the Swift River.

Gold is about seven times as dense as ordinary sand grains or pebbles, so a current strong enough to move a small piece of gold can move a much larger piece of common rock. For this reason, gold tends to accumulate in gravel and coarse sand deposits. A gold pan duplicates the stream's method of concentrating the gold.

Fill half the gold pan with stream gravel and top it off with water. Shake the pan sideways to settle the heavier minerals, then swirl and tilt the pan to wash out the upper layers of sand and gravel. Any gold will already have sunk to the bottom. Continue the process with more gravel and more water, removing the pebbles by washing or picking them out. As you continue panning, the residue in the pan will become red with garnets, below which it is black with magnetite, the magnetic oxide of iron. Any gold will be hiding in the black sand. If you find no gold, the finding of which requires patience and luck, the best thing in the pan will be the abundant red garnet.

The best panning is along the East Branch Swift River, a little more than a mile and a half from Coos Canyon on the road to the east that crosses the Swift River and leads past Tumbledown Mountain. Take the left fork in the road. The panning area is where the stream comes back to the road. This placer is preserved as a recreational gold deposit and no power equipment is allowed.

The rock in which the beautiful potholes were eroded at Coos Canyon is the Perry Mountain formation, which was deposited as mud during Silurian time. Now it is a metamorphic rock, basically a mica schist that also contains staurolite and garnet. Look closely at the fresh rock to see the little black crosses of staurolite, twinned crystals. The garnets are red.

Rumford Glacial Erratic

A house-size boulder of granite sits right beside U.S. 2 as it passes around Rumford. The bedrock here is the Seboomook formation, originally mud and sand deposited on the ocean floor. The granite block, along with its smaller neighbor, may have been carried by the last ice sheet from one of the small granite bodies near Black Mountain or one of the closer hills.

Black Mountain Pegmatite

Black Mountain is a few miles northwest of Rumford, but it must be approached from the north, from Maine 120. The pegmatite quarry

Erratic boulders of granite in Rumford. It is unusual for such large rock fragments to move very far from their bedrock source before the glacier decides to drop them.

is on the west side of Black Mountain, near the summit. It is chiefly noted for its pink and purple mica called lepidolite—absolutely beautiful stuff, world-class. Lovely crystals of pink tourmaline are intergrown in the lepidolite. And it contains long white or pinkish blades of spodumene, a rare kind of pyroxene. Lepidolite and spodumene are lithium silicates, the main sources of this rare and useful element. The famous Newry gem mines are across the Ellis River valley to the west. The pegmatite quarry is open to collectors only with the permission of the owners.

Newry Mines

At least ten separate mines have worked the Newry pegmatite deposit. They are on the eastern slopes of Plumbago Mountain, about 2 miles west of Maine 5. They are open to mineral collectors for a fee.

The Dunton mine is most famous for its beautiful crystals of watermelon tourmaline, green on the outside and pink in the middle. A number of jewelry stores produced tourmaline from pits in this vicinity in the late 1800s.

In 1972, a local man took up mineral collecting after being advised to get more exercise. Someone advised him to try the old Dunton

Spodumene crystal (under knife) in lepidolite mica on Black Mountain, Rumford.

pegmatite. Still knowing nothing about minerals, he started to bang on the solid pegmatite with his new rock hammer. The wall suddenly gave way, exposing an open cavity, from which he extracted some large tourmaline crystals of gem quality. He showed them to a rockhound on the way home, who told him he was holding a small fortune. Thus began a minor boom in tourmaline mining. The specimens from the Newry mines were at least partly the reason that tourmaline was named the state mineral. More than eighty minerals have been identified from this mine.

The Nevel mine mainly produced spodumene for its lithium content. You can see crystals 20 to 30 feet long exposed in the mine face, especially over the twin tunnels. About forty other minerals have been found. The Bell pit is noted for its wide variety of rare phosphate minerals.

Andover Basin

From Black Mountain and several places near the Newry mines, you can see that Andover lies within a semicircular basin. When I first saw it, I hoped it might be a meteorite impact crater. The facts are much less exciting.

The mountains surrounding the basin are mostly hornfels, or baked slaty sedimentary rocks. The basin was eroded into the granite pluton that baked the hornfels when it intruded as molten magma during Devonian time. Hornfels resists weathering and erosion better than granite, so it stands as a ridge and the granite erodes into a basin. The granite is the southern extension of the Mooselookmeguntic granite.

Set in the middle of this basin are a number of large satellite communication dishes, an early type known as Telstar. The basin site was chosen because the surrounding rim of mountains reduced electrical interference.

Bear River Gorges

A number of interesting things exist near where Maine 26 crosses the Mahoosuc Range and the Bear River. Upstream from the Sunday River ski area in Grafton, the river flows through three narrow gorges: Screw Auger Falls, named for its tightly meandering route; Mother Walker Falls; and Moose Cave Falls. All formed as lines of giant potholes ground into Devonian igneous rocks—granite and gabbro.

Potholes form where persistent eddies or whirlpools turn rocks on the bedrock floor of a stream, slowly grinding a more or less cylindrical hole. These are oversized. Even in flood, the Bear River does not carry enough water to grind such large potholes. They probably formed

Screw Auger Falls on Bear River, Grafton.

at the end of the last ice age, when a melting glacier overlooked the head of the valley and shed torrents of water and sediment down it.

Downstream from these gorges the Bear River flows through a plain of glacial outwash that fills the valley.

Table Rock and Slab Caves

Table Rock is a large, flat boulder of granite that lies on the mountainside above the head of Bear Valley; it gives a fine view of the valley below and the Mahoosuc Range to the west. Slabs of granite similar to Table Rock make caves, of a sort, on the slope below.

Granite that was buried deep in the earth expands slightly as erosion removes the burden of rock that once covered it. The expansion opens cracks called sheeting joints, which are parallel to the eroding land surface. In this case, they tilt down toward the valley floor. Large slabs of granite have broken from the sheeting joints above and slid down the slope, forming a jumble below. The spaces between them are slab caves, the largest in Maine.

Snow and ice accumulate in the deeper parts of the slab caves and in some years last well into the summer. They would make slab cave exploration a pleasant pastime on a hot summer day, were it not for all the porcupine dung. A flashlight and old clothes are advised for exploring caves like this.

Old Speck Mountain

Old Speck Mountain, at 4,180 feet, is the third highest in Maine, the highest point on the Appalachian Trail in Maine until you reach Mt. Katahdin, more than 150 miles to the northeast. Several trails reach the summit, which provides a full panorama of the mountains of Maine and New Hampshire. Old Speck also has some nice glacial cirques, deep basins carved by small valley glaciers. The bedrock is the Silurian Rangeley formation, near its contact with Devonian granite.

Androscoggin River

The Androscoggin River is the third largest in Maine; it drains an area of more than 3,500 square miles. Its course begins in Maine, near Rangeley, passes through New Hampshire between Errol and Gorham, and finally returns to Maine. Between Gorham and Rumford in Maine, it flows northeast, parallel to the bedrock structure, before turning south for the sea. South of Bethel are several large valleys, now filled with sandy glacial deposits, that lead directly to the Atlantic near Portland. It may well be that the Androscoggin River flowed through one of these before the ice ages, developing its present course after glacial deposits plugged the old valley.

Bethel

This attractive town has great views of the White Mountains and the Mahoosuc Range. From the early days of World War II until the 1960s, all Maine mica was sorted and sold in Bethel. Then transistors replaced vacuum tubes, and that put Bethel out of the mica business.

Evans Notch

Maine 113, the road through Evans Notch south of Gilead, affords marvelous views of the White Mountains. The entire length of this road, which is closed in winter, lies in the White Mountain National Forest. A spectacular waterfall, named Bickford Slides, is about 1.5 miles from the Evans Notch Road on the Speckled Mountain Trail. The falls tumble across a glacially eroded outcrop of the Smalls Falls formation, at this locality gneiss. The relic bedding is nearly horizontal. Speckled Mountain is named for the glinting mica flakes in the schistose rocks of the Perry Mountain and Smalls Falls formations at its summit.

Maine 4
Farmington—Rangeley
42 miles

Before World War II, Farmington was the terminus of quite a remarkable system of narrow-gauge railroads known as the "two-footers," referring to the width between the tracks. At Farmington a standard-gauge line, with tracks 4 feet, 8.5 inches apart, ended where the two-footers began, and freight and passengers were exchanged. The line was the Sandy River and Rangeley Railroad. Some of its rolling stock was moved to the Edaville Railroad in Carver, Massachusetts. More recently, most of the engines and cars were moved to Portland, where the train runs as a tourist attraction.

The Farmington Fair, officially known as the Franklin County Agricultural Exhibition, remains one of the few old-fashioned country fairs, with few commercial intrusions.

This segment of Maine 4 crosses Devonian schists and sandstones in the southern part, generally coarser Silurian sediments in the central part, and Ordovician turbidite mudstones and volcanic rocks near Rangeley. The Devonian rocks are thought to be metamorphosed

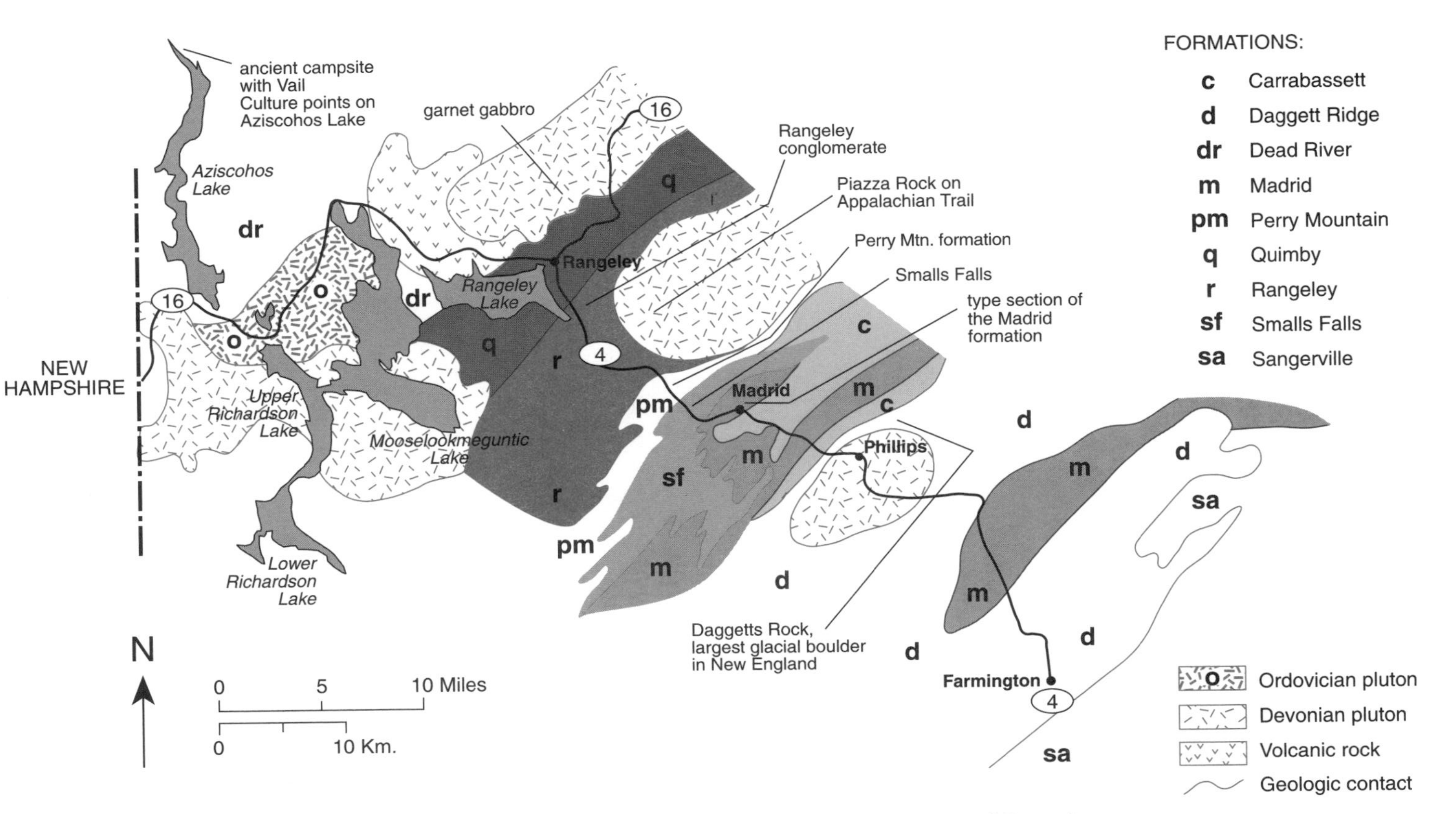

Geologic features along Maine 4 between Farmington and Rangeley.

versions of the Seboomook group of formations that were deposited in deep water as turbidite muds. The Silurian sediments appear to have been laid down on the deep ocean floor, at the mouths of submarine canyons closer to the continental shelf. The Ordovician turbidites and volcanic rocks are part of the Bronson Hill terrane, a chain of volcanoes that collided with North America during Ordovician time.

Much of the road follows close to the Sandy River, a western tributary of the Kennebec River. The Rangeley Lakes are part of the headwaters of the Androscoggin River.

Carrabassett Formation

The Carrabassett formation is the youngest in this area. It was deposited as muddy sediments during early Devonian time, then metamorphosed during the Acadian mountain-building event. The formation consists of thin beds that are sandy near the base and grade upward into fine mud. Now that they are metamorphosed, the muddy parts are staurolite schist. Staurolite is the mineral that generally crystallizes into little crosses.

Daggetts Rock

Daggetts Rock, northeast of Phillips off Wheeler Hill Road, may be the world's largest glacially transported boulder. This piece of granite was probably broken from outcrops on the hill above, and the glacier decided that this was far enough to take it. It is also possible that the huge block simply broke from the cliff and rolled or slid down the hill. Daggetts Rock measures about 100 feet by 30 feet by 20 feet, and weighs about 5,000 tons. It is cracked down the middle.

Phillips was the hub of a system of narrow-gauge railroads that fanned out to several towns and logging operations to the north and northwest. The standard-gauge rails came as far as Farmington, where goods and passengers transferred to the two-footers. Much of the track and rolling stock was sold before World War II.

Madrid Formation

The Madrid formation is named for a small village on the Sandy River; pronounce it with emphasis on the first syllable. At Madrid, near the mouth of Saddleback Stream, an upper section of thinly bedded turbidite mudstones grades upward into thicker sandstone layers sandwiched between thin layers of mudstone. The mudstone layers contain crystals of staurolite—little crosses. Along the Sandy River, the Madrid formation consists of thick sandy beds. Most of the Madrid formation exposed to the northeast of the type locality also consists of thick sandy beds and is much less metamorphosed than the rock at Madrid village.

Many outcrops of the Madrid formation contain nodules shaped like footballs that were originally concretions of calcium carbonate—limestone. Metamorphism has converted these pods to masses of calcium silicate minerals. Look for crystals of dull green diopside and reddish brown garnet.

The original sediments of the Madrid formation and younger rocks like the Carrabassett formation appear to have come from the east, presumably from the Avalon terrane, which was then approaching North America. It finally collided during the Acadian mountain-building event of Devonian time.

All the older sediments in this part of Maine came from the northwest, from North America. In particular, many of them came from the mountains that rose during the Taconic mountain-building event of Ordovician time. Rocks exposed farther north on this road are older than the Madrid formation. They contain sand ripples and other current features that indicate a source to the northwest.

Smalls Falls

A beautiful series of waterfalls and potholes on the Sandy River and its tributary, Chandler Mill Brook, exposes a thinly bedded turbidite mudstone named the Smalls Falls formation. It weathers to a rusty color. Geologists think it was deposited during late Silurian time.

The Smalls Falls mudstone is a dark rock that contains thin blebs of the mineral pyrrhotite, a somewhat unstable compound of iron and sulfur. On exposure to the atmosphere, the iron dissolves as the sulfur combines with oxygen and water to form sulfuric acid. The dissolved iron then combines with atmospheric oxygen to form the rusty coating of iron oxide that stains these outcrops. A close sniff of a freshly broken piece of the Smalls Falls mudstone also reveals the unmistakable rotten-egg odor of hydrogen sulfide.

Pyrrhotite is weakly magnetic, just enough to throw compasses a bit off if you get them too close to the rock. A lightning strike strengthens the magnetism of almost any rock, commonly enough to turn a compass needle around. While I was mapping the rocks in another part of the state, I found a ledge that made the compass point the wrong way entirely. When I asked the landowner if lightning had recently struck that ledge, she replied, "No, but a UFO landed back there last Saturday night." Evidently they have the same magnetic effect on rocks that lightning does.

The best place to see the graded beds in the Smalls Falls formation is at the top of a waterfall on Chandler Brook, a few hundred feet west of Smalls Falls on a well-worn trail. On Maine 4, exposures of the rusty rocks of the Smalls Falls formation give way to the underlying Perry Mountain formation, a brilliant white turbidite mudstone with thick layers. It is a metamorphic rock that contains crystals of staurolite, a mineral that characteristically crystallizes into little crosses. The best exposure here, one that is known on geology field trips as the most dangerous outcrop in Maine, is on the inside of a sharp bend in the road.

The difference between the Smalls Falls and Perry Mountain formations probably reflects conditions on the seafloor when they were deposited. Oxygen decomposes organic matter in sediments. If no oxygen is available, the organic matter survives and dark sediments like the Smalls Falls mudstone accumulate. Those tend to contain sulfide minerals such as pyrrhotite. Where abundant oxygen is available the organic matter decomposes, leaving white quartz as the main component of the sediment as in the Perry Mountain formation. Placer gold is said to occur in the stream gravels at Smalls Falls, but I have never been lucky enough to find any here.

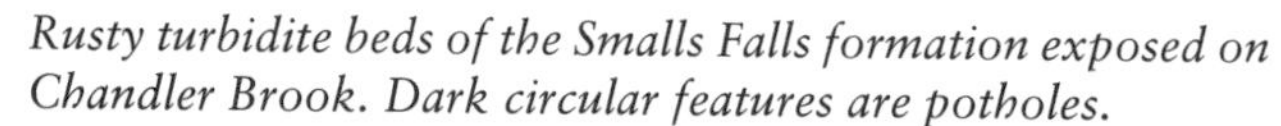

Rusty turbidite beds of the Smalls Falls formation exposed on Chandler Brook. Dark circular features are potholes.

Inclined beds of the Perry Mountain formation on Maine 4, Township E.

Piazza Rock

The Appalachian Trail crosses Maine 4 a couple of miles from Smalls Falls. To the northeast, the trail passes from Silurian sedimentary rocks to the Redington granite pluton.

Like many granites, the Redington pluton tends to break into slabs along fractures more or less parallel to the ground surface. One such slab slid to an overhanging position, forming Piazza Rock. In old New England parlance, a piazza is a porch or veranda, especially a covered one. This forested slab must have reminded the person who named it of a porch. It resembles Table Rock, near Old Speck Mountain, also just off the Appalachian Trail.

Several slab caves are near Piazza Rock. They are simply openings between granite slabs that piled up, helter-skelter, near the bottom of the cliff. They normally preserve snow and ice well into the summer.

Rangeley Formation

The Rangeley formation is a pebbly sandstone that was deposited during early Silurian time. The pebbles match Precambrian, Cambrian, and Ordovician rocks now exposed farther north. The source area was probably mountainous during early Silurian time.

Many of the pebbles in the Rangeley formation are elongated and penetrate one another. They probably dissolved at points of contact, where the compression of the Acadian mountain-building event put

Stretched pebbles of a variety of rock types in the Rangeley A conglomerate, Greenvale Cove.

them under great pressure. The mineral quartz is especially vulnerable to pressure solution; it dissolves where pressure is great, then precipitates in fractures, where pressure is low, to make quartz veins.

As the Rangeley formation was deposited, the mountain range that was its source was eroding. The pebbles in the conglomerates become smaller upward, perhaps because slopes in the source area were becoming flatter. The last conglomerate beds deposited contain only quartz pebbles derived from quartz veins, commonly the rock that best resists weathering and erosion. Many years after mountains are formed, all that remains are quartz pebbles.

The lower part of this conglomerate is called the Rangeley A conglomerate. Look for it along Maine 4 near the southeast end of Rangeley Lake in Greenvale Cove. The Rangeley A conglomerate contains pebbles of a variety of rocks, many of which are similar to rocks now exposed farther north. Some of the pebbles contain bright blue quartz grains like those in the Chain Lakes gneiss near the Canadian border.

The blue quartz contains thin needles of rutile, or titanium oxide. They scatter light rays to give the quartz its blue color. In other places, much larger rutile needles grow in quartz, making a truly beautiful stone.

The Rangeley B conglomerate is nicely exposed along a trail that starts at the former Greenvale Cove School, now the town hall, and follows Cascade Stream. The lower part contains small quartz pebbles, and the upper part is a turbidite mudstone. The mudstone is metamorphosed to schist, which contains exquisite little crosses of staurolite. Another mile along this trail brings you to Cascade Stream Gorge, eroded in the same rocks. It is one of the most spectacular gorges in Maine, as much as 100 feet deep and nearly a half-mile long.

Rangeley

The town and the lake were named for the original settler, James Rangeley, who moved here in 1825. With the establishment of the Sandy River and Rangeley Railroad in the 1850s, this area became one of the earliest inland summer resorts. The lakes in the Rangeley area are known for their fishing and for the stable Rangeley boat, a sort of wide canoe with square stern, made of cedar. A quarry near the Rangeley airport has worked an igneous rock for its garnet that makes excellent sandpaper.

Saddleback Mountain

Bedrock on the summit of Saddleback Mountain is a scrap of the Smalls Falls and Perry Mountain formations that was surrounded by the magma of the Redington pluton and baked to a hornfels in its heat. That happened during Devonian time, perhaps about 380 million years ago.

Hornfels better resists weathering and erosion than any other kind of bedrock that exists in significant volume in Maine. The range of mountains to the east stands in bold erosional relief because their bedrock is hornfels. The peaks include Mt. Abraham, Sugarloaf Mountain, and Bigelow Mountain.

Orbeton Stream and the Dallas Esker

Orbeton Stream flows from Reddington Pond down the steep valley between Saddleback Mountain and Mt. Abraham to the Sandy River near Phillips. Much of its course crosses the Redington granite pluton that invaded the older rocks in this vicinity during Devonian time, while the Acadian mountain-building event was in progress.

Orbeton Stream follows the course of a major meltwater stream that flowed here as the ice melted at the end of the last ice age. That vanished torrent left its souvenirs in an esker that follows Orbeton Stream and ends in a large boulder fan near Toothaker Pond.

One of the equally vanished narrow-gauge railroad lines also followed this valley between Phillips and Dallas. The old roadbed made a good trail for many years after the rails were sold for scrap before World War II, especially for at least two geologists who used it in the 1970s. In later years the Navy has gated the road to protect some sort of survival training program somewhere near Redington Pond. Local opinion is that working there during black fly season toughens the officers.

Aziscohos and Richardson Lakes and an Ancient Campsite

Aziscohos and Richardson Lakes have a long, narrow meandering shape that strongly suggests river valleys. The Androscoggin River probably flowed here before the ice ages, on its way south toward the coast near Portland. Glaciers eroded the western sides of the river valleys, then dumped sediment into them as they melted, converting them into lake basins. Dams now regulate all these lakes, and Mooselookmeguntic Lake to the east, to control stream flow for power plants downstream.

An ancient campsite was found near the mouth of the Megallaway River when Aziscohos Lake was lowered for dam repair. Radiocarbon dates on charcoal from campfires, perhaps mixed with charcoal from forest fires, gave ages between 10,300 and 11,120 years ago, about at the end of the last ice age. The site yielded beautifully worked fluted points typical of the Vail Culture. Unlike other ancient campsites in the northeast, the Aziscohos site contained no caribou remains. Although other sites of the Vail Culture exist in Maine, the one at Aziscohos Lake is the oldest with radiocarbon dates to substantiate its antiquity.

At the end of the last ice age, Maine was covered with tundra vegetation and small trees, which made the interior relatively accessible. When forests returned, human occupation of Maine ceased for many years, except along the coast and the major rivers.

Maine 27
Farmington—Coburn Gore
74 miles

Maine 27 traverses a great thickness of sedimentary rocks deposited at the edge of the North American continent during Silurian and Devonian time. Many valleys in the area have been formed in plutonic rock, especially gabbros and diorites. One valley passes between two of Maine's highest mountains, Sugarloaf and Bigelow, which were created by the heat of the Sugarloaf gabbro baking muddy sediments of the Carrabassett formation on Bigelow and similar Silurian rocks on Sugarloaf. Between Stratton and the Canadian border at Coburn Gore is the Boil Mountain ophiolite and the mysterious Chain Lakes massif.

New Vineyard End Moraine

About 9 miles after leaving Maine 4, Maine 27 climbs to a narrow pass between the mountains on either side. During its retreat, the last glacier stood here long enough to build a hummocky moraine with numerous boulders. Meltwater from this ice carried the outwash that fills the valley of Barkers Stream as far south as Fairbanks.

Years ago I drilled through this outwash, into clay of the Presumpscot formation. On penetrating the clay, the auger hit some gravel. When the auger was pulled out of the hole, groundwater under artesian pressure squirted about 5 feet into the air. This spring continued to flow for many years, visible from Maine 27, until some local boys plugged the hole with sticks.

New Portland and the Wire Bridge

New Portland was given to the people of Portland by Massachusetts, to repay them after part of their town was burned by the British fleet in 1775. Crossing the Carrabassett River at New Portland is a bridge that must be one of the engineering marvels of the world. Built in 1846, it is supported by wire, actually fine English steel wire braided into thin cables. The cables were brought here from the Kennebec River by a wagon pulled by sixteen yoke of oxen. The wooden timbers and boards are suspended from the cables by thinner strands of wire. It sways and creaks a little but is still open for light traffic. It was long known as Colonel Morse's Fool Bridge, after the designer. The bridge has been placed on the Register of National Historic Sites. Small exposures of the Madrid formation occur along the stream bank near the bridge.

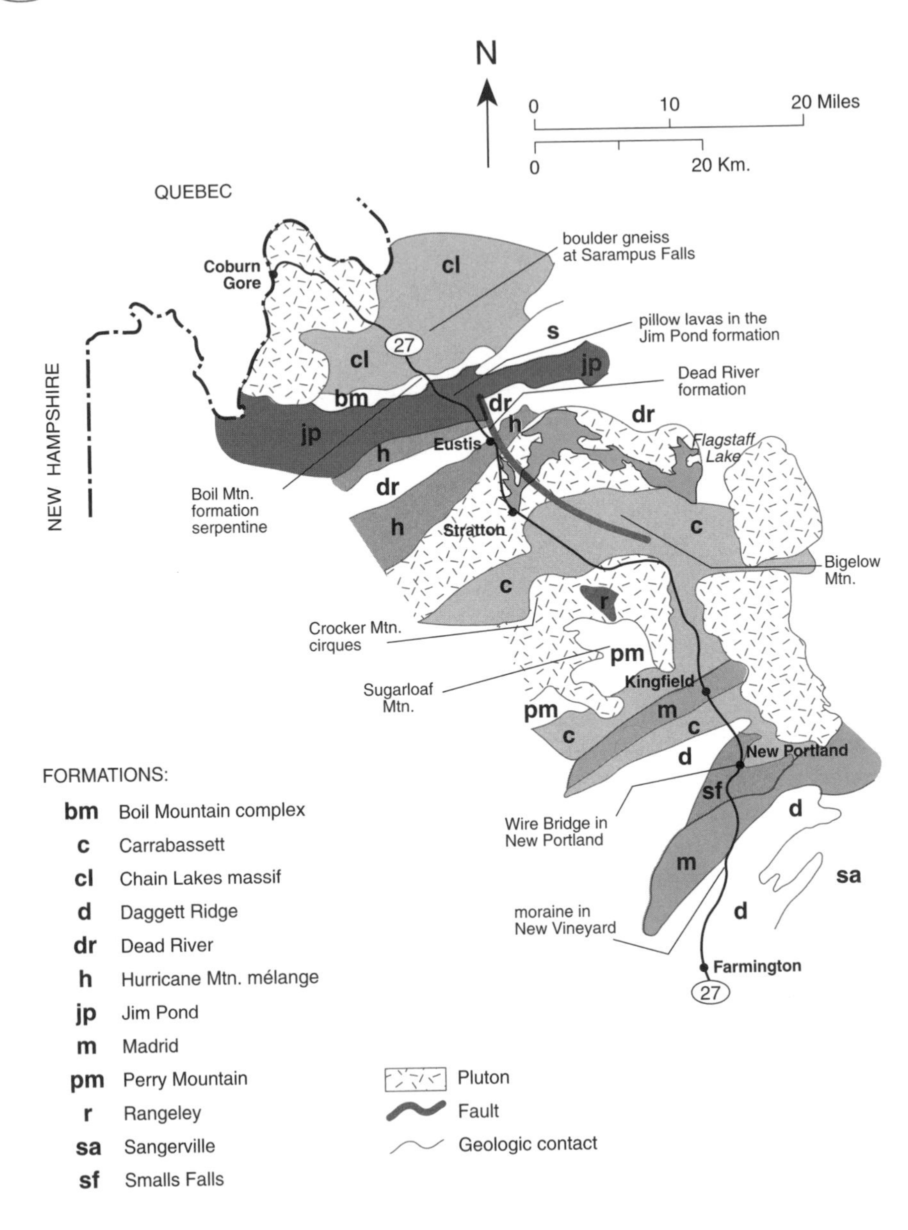

Geologic features along Maine 27 from Farmington to Coburn Gore.

Kingfield

Kingfield has long been associated with logging, and still has a large wood products mill. The town has also profited, some would say, from the skiers at Sugarloaf Mountain. Kingfield is famous as the birthplace of the Stanley brothers, who invented a steam-driven automobile that in its day was the fastest thing on wheels. The Herbert Hotel here was built by a lumber baron so he would have a suitable place in which to entertain his rich friends from Portland and Boston.

Sugarloaf and Bigelow Mountains

In 1775, Major George Bigelow was a member of General Benedict Arnold's expedition against the British in Québec. When he followed the Dead River to this area, Arnold sent Bigelow up the mountain now named for him, to determine if he could see Québec City, about 100 miles north of here. He could not.

Sugarloaf and Bigelow Mountains are eroded in slaty rocks that were baked in the heat of the Sugarloaf gabbro during Devonian time. The hornfels is resistant to weathering and erosion, and that explains why these mountains stand so high. The hornfels zone on Bigelow is quite thin, which may partly explain the narrow ridge along the crest of the mountain. Sugarloaf Mountain is the second highest in Maine;

Wire Bridge over Carrabassett River in New Portland.

Bigelow is the fourth highest. For many years, the elevation of Sugarloaf was listed as 5,240 feet. Actually it is about 4,237 feet. The Appalachian Trail formerly crossed Sugarloaf but was moved north to Crocker Mountain to avoid the development associated with the ski area on Sugarloaf.

The summits of Sugarloaf and Bigelow Mountains are above tree line; they maintain small communities of tundra plants, a relic of the tundras that existed nearly everywhere in New England at the end of the last ice age. Relict tundra also covers a large area of Katahdin's Table Land and much of the Presidential Range in New Hampshire.

Investigation of unstable supports on the ski gondola on Sugarloaf Mountain revealed permafrost—permanently frozen ground. Permanently frozen ground occurs where the annual temperature is below freezing, a condition that the tundra plants also require. Permafrost probably exists on Katahdin as well.

Crocker Mountain Twin Cirques

Small valley glaciers eroded two glacial cirques into the east side of Crocker Mountain. They are notable because their floors are lower than those of other cirques in Maine. Cirque floor elevations roughly mark the lowest point of the permanent snow needed to supply the glacier. The floors of the Crocker cirques are about 1,800 feet above sea level, compared with about 3,000 feet on Katahdin.

The elevation of the snow line is determined by the annual snowfall and by the amount of snow that melts in the summer. The Crocker Mountain cirques face eastward, as do several of those on Katahdin, so both should have about the same amount of summer melting. It would seem then that the Crocker Mountain area got more snow than Katahdin did when these cirques were formed. I know of no studies that show that Crocker Mountain gets more snow than Mt. Katahdin today.

Flagstaff Lake and Glacial Lake Bigelow

Flagstaff Plantation was named for the site where General Benedict Arnold and his troops camped for three days, with the Continental flag flying from a tall pole outside the leader's tent. At the time of Arnold's expedition the Dead River flowed through this valley. The dam at Long Falls was built in 1950 to impound Flagstaff Lake, which flooded the village of Flagstaff.

During the retreat of the last ice sheet, Glacial Lake Bigelow filled the valley that Flagstaff Lake now occupies. A set of deltas near Kennebago Lake records several stages of this lake, each lower than the one before. During the highest stage, the lake drained south through Dallas Plantation to the Sandy River by way of Orbeton Stream. When the ice margin melted back north of Stratton, the valley between Sugarloaf and Bigelow Mountains was the drainage for the next lowest lake stage. When the ice margin pulled back from the north slopes of Bigelow, the lake found a lower outlet around the south end of Bigelow and back into the Carrabassett River by way of Poplar Stream.

Outflow through Poplar Stream deposited a large boulder fan in the river valley at the town of Carrabassett Valley. The Carrabassett River still detours around the margin of the fan near the Carrabassett Valley airfield. Glacial Lake Bigelow finally drained when the ice front melted back from the Dead River valley, the present outlet of Flagstaff Lake.

Eustis Ridge and the Cathedral Pines

The Cathedral Pines between Stratton and Eustis grow on a sandy delta deposited in a low stage of Glacial Lake Bigelow. These tall red pines are spectacular, even if not virgin timber.

The picnic area on Eustis Ridge provides a good panorama. You can see a large part of Flagstaff Lake in the foreground, with Bigelow Mountain to the right. The tall stack at the west end of the Bigelow Range, in Stratton, is an electrical generating plant that burns wood. Farther to the right is Sugarloaf, where snow usually lasts well into the summer.

Dead River Formation

Watch just north of the village of Eustis for a small outcrop of greenish slaty rock, part of the Dead River formation. The original sediments were deposited during Ordovician time, then squashed during the Taconic and Acadian mountain-building events. That double deformation left the rock with slaty cleavage in two intersecting directions. Now it tends to break into long splinters and rhombohedral blocks, instead of into the slabs typical of a simple slate that cleaves in only one direction.

Hurricane Mountain Mélange

The Hurricane Mountain mélange outcrops in a narrow belt that extends more than 90 miles northeast of Eustis. The rock consists of a chaotic mess of blocks of gabbro, basalt, sandstone, andesite, and chert set in a matrix of slate. Rocks of this type typically form as muddy sediments scraped off oceanic crust sinking through an oceanic trench, and mix with whatever other rocks may be available. In most outcrops you see only the exotic blocks because the matrix so easily weathers and erodes. The exotic blocks make rounded knobs called knockers, for their resemblance to the weight in a bell that does the ringing.

The slaty part of the formation contains fossil sponge spicules that indicate a Cambrian age. Then it was scraped together somewhere far to the east of its present position and left here during the Taconic mountain-building event of Ordovician time.

Boil Mountain Ophiolite

Excellent exposures of the Boil Mountain ophiolite appear along Maine 27 in the 4 miles between Alder Brook and Sarampus Falls, on the North Branch Dead River. Ophiolites are scraps of oceanic crust that were shoved onto a continent. This bit of old ocean floor dates from Cambrian time. It was in the floor of the Iapetus Ocean.

The Boil Mountain ophiolite was shoved onto the Chain Lakes massif in early Ordovician time during the Penobscot mountain-building event, which long preceded the Taconic mountain-building event. The originally horizontal basalt lava flows now stand on edge.

As you drive north along Maine 27, you pass through the old oceanic crust from the top down. The exposed rocks are pillow basalts of the Jim Pond formation, pinkish gray plagiogranite, and greenish serpentinite.

Jim Pond Formation

The Jim Pond formation consists of pillow basalt—lava flows that erupted on the deep ocean floor. Basalt pillows form when streams of molten lava erupt into cold seawater; the outer part of the flow chills against the water, while the inner part remains molten. Lava bursts through the hard outer skin and pours out onto the ocean floor in a long stream, the outer part of which promptly chills. As that happens again and again, the lava flow becomes a pile of long cylinders of basalt. When you see them sectioned in a roadcut, they look like a pile of greenish black pillows.

While they are still soft, the cylinders of basalt shape themselves to the older cylinders beneath. When you see them in a roadcut, the bottom of each pillow looks as though it had been pinched, while the top is rounded. The pillows exposed on Maine 27 are on end, with their rounded tops, which originally faced up, now facing south. Jasper, a bright red variety of the quartz mineral chert, fills the spaces between some of the pillows. The basalt also contains widely scattered crystals of the copper mineral chalcopyrite; they look metallic and have a rich golden color.

Pillow basalt of the Jim Pond formation in Alder Stream Township.

Plagiogranite

Plagiogranite, or trondhjemite, is an unusual rock that occurs in ophiolites. Normal granite typically contains quite a lot of orthoclase feldspar, a much smaller proportion of plagioclase feldspar, some quartz, and a speckling of black minerals, such as biotite mica. Plagiogranites consist largely of plagioclase feldspar and quartz. Many also contain ore minerals. Prospectors looking for gold dug a prospect pit in pinkish and rusty plagiogranite just east of Maine 27. I have no idea what they found—obviously not enough to inspire anyone to invest in a mining operation.

Serpentinite

Serpentinite is a curious rock composed largely of the curious mineral serpentine. Serpentinite comes in various shades of green, yellowish green, and black. It feels a bit like soap, so most people call it soapstone. Everyone has seen serpentinite carved into little gift-shop figurines.

Serpentinite forms when the black peridotite of the earth's mantle reacts with hot water at the rift in an oceanic ridge, at the boundary between separating lithospheric plates. So it is no surprise to find serpentinite at the base of the Boil Mountain ophiolite, a situation in which peridotite is common. The lowest part of the oceanic crust is peridotite, and ophiolites are slabs of oceanic crust.

Most serpentinites contain asbestos, and some occurs here. In the Thetford mines area of Québec, about 40 miles north of Jim Pond Township, is another, much larger, ophiolite where asbestos has been mined for years. The Thetford ophiolite was shoved onto North America during the Taconic mountain-building event of Ordovician time.

A Long View in Time from the Chain Lakes Massif

The boulder gneiss phase of the Chain Lakes massif is nicely exposed at Sarampus Falls. Near a highway garage are outcrops of the second phase of the massif, the Flecky gneiss. These rocks are believed to be about 1.6 billion years old, although some believe it is Grenville crust, about 1.0 billion years old. In one outcrop of the boulder gneiss, the gneiss contains cobbles that have a metamorphic fabric that is inclined to the fabric in the gneiss at an angle of about 40 degrees. Why are the metamorphic fabrics in the gneiss and the cobbles not parallel?

One explanation has the cobbles eroding from some ancient gneiss even older than the Chain Lakes massif. They were deposited in a

Boulder gneiss of the Chain Lakes massif, near Sarampus Falls in Alder Stream Township.

sediment that was later metamorphosed. Another explanation has the cobbles as debris from an asteroid impact explosion.

From the beach at the north end of Natantis Pond, you can look down the valley of the North Branch Dead River all the way to Bigelow Mountain, 10 miles away. Bedrock there is the early Devonian Carrabassett formation.

Between Natantis Pond and Bigelow Mountain, the bedrock consists of the Chain Lakes massif, the Boil Mountain ophiolite, and some Cambrian and Ordovician rocks not well exposed along Maine 27. All of these rocks are essentially stacked on end. The view from the 1.6-billion-year-old Chain Lakes rocks to the 400-million-year-old Carrabassett formation spans an abyss of time comparable to that you see looking up from the bottom of the Grand Canyon.

Chain Lakes Esker

Along much of Maine 27 between the border at Coburn Gore to Eustis, the highway is on or beside an esker. It was part of the glacial meltwater system that carried sand and gravel to the deltas at the northwest end of Glacial Lake Bigelow.

It is interesting to note that in its final stages in northern Maine, the last ice sheet was reduced to an ice cap centered near the Boundary Mountains. From this center, ice flowed both south into Maine and north into the St. Lawrence River valley.

U.S. 201
The Arnold Trail
Skowhegan—Canadian Border

88 miles

Benedict Arnold left Cambridge, Massachusetts, on September 13, 1775, and arrived at the mouth of the Kennebec River on September 22. With eleven hundred American troops, several camp followers, and two hundred bateaux loaded with food and supplies, he embarked from Pittston, south of Augusta, on a grueling wilderness campaign to attack the British in Québec City. The expedition covered some 600 miles in eight weeks. Only five hundred fatigued and starving troops made the final destination; many of those perished or were captured during the attack on Québec City.

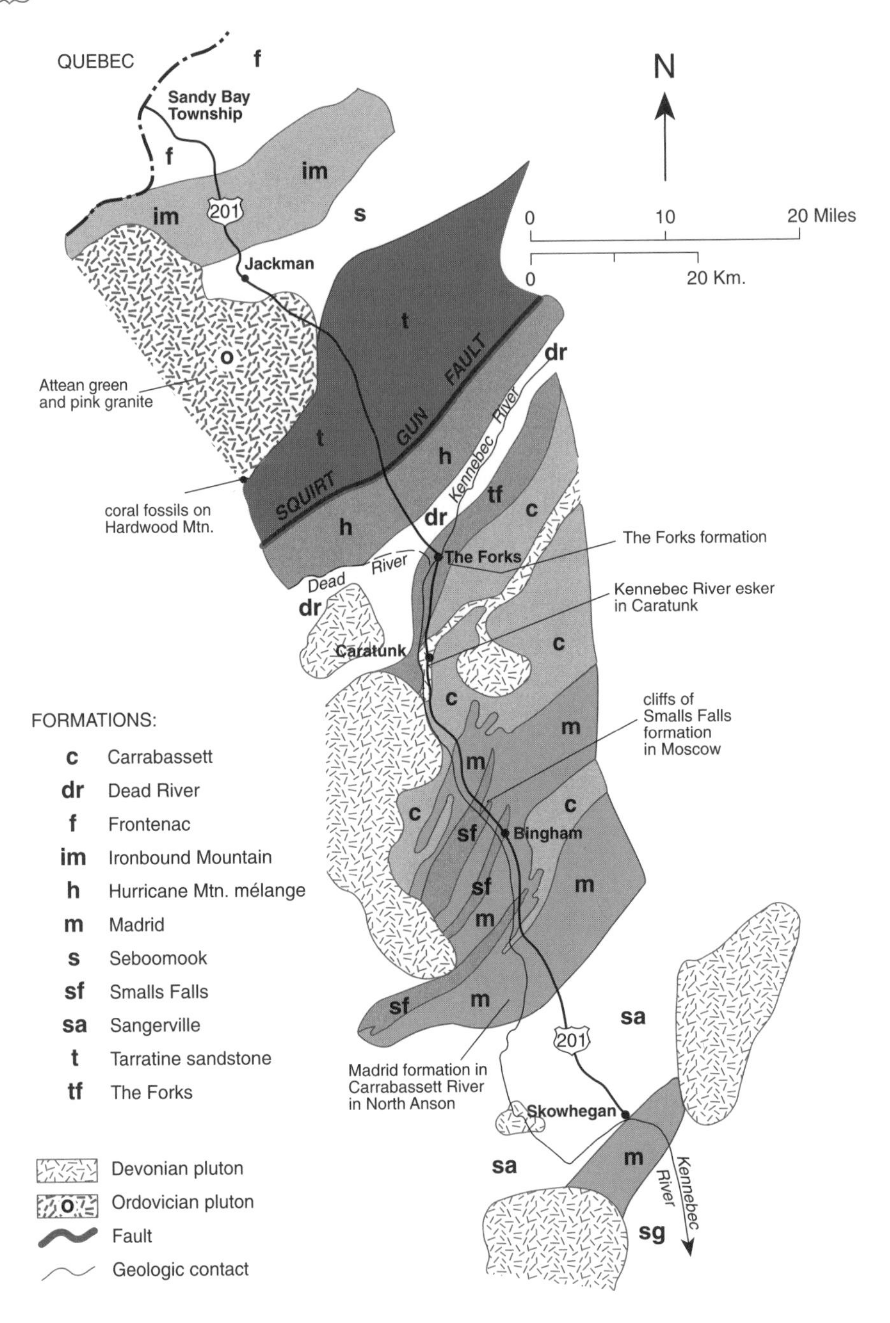

Geologic features along U.S. 201 from Skowhegan to the Québec border at Sandy Bay Township.

This part of U.S. 201, the Arnold Trail, follows the Kennebec River and traces the first leg of Arnold's journey by boat across the perilous Maine wilderness. Although he and his men were certainly not geologists, you can safely assume that they described every ledge and falls in their diaries.

Between Waterville and The Forks, U.S. 201 and the Kennebec River traverse the northwest edge of the Central Maine slate belt. Some of the best exposures of the Silurian Waterville formation in Waterville, Sangerville formation in Skowhegan, Smalls Falls formation from Bingham to Caratunk, and Madrid formation in Solon and North Anson can be seen along this route. Most of these outcrops are in large ledges along the Kennebec River that proved disastrous for the Arnold expedition.

North of The Forks, the road crosses the Lobster Mountain anticline. The anticline raises to the surface Cambrian and Ordovician rocks that were crumpled into folds during the Taconic mountain-building event of Ordovician time. Bedrock between Parlin Pond and Jackman is Devonian sandstone, the Tarratine formation, with minor outcrops of the overlying Tomhegan formation, a Devonian turbidite. These are some of the youngest rocks in this part of Maine. The Attean pluton, a mass of granite emplaced during Ordovician time, fills a large area west of Jackman. Between Jackman and the border are various mudstones that were deposited in deep water during Ordovician, Silurian, and Devonian time. The Boundary Mountain anticline raised them to the surface.

The Arnold Trail

Bedrock between Waterville and Skowhegan is mostly the Waterville and Sangerville formations. But thick deposits of glacial sands, marine clays, and river silts cover most of it. These deposits fill the Kennebec Valley between Waterville and Bingham, creating a fertile soil that the Indians cultivated before the colonial settlers arrived. The first good exposure north of Waterville is the Sangerville formation. It appears to be the same as the Perry Mountain formation and Smalls Falls formation in Skowhegan.

Between Norridgewock and Skowhegan, the Kennebec River flows northeast, following the regional grain of the bedrock. At Skowhegan, the river flows through a narrow gorge and abruptly turns back to its normal southeasterly course, creating the Great Eddy. It tossed many of Benedict Arnold's bateaux against the rocks. The survivors were carried up the vertical walls at the base of the falls. A dam and hydroelectric plant now occupy the site, as is the case with many of the falls and gorges along the major rivers in Maine. Eleven major dams block the Kennebec River.

The black Smalls Falls formation with its gross overabundance of iron sulfide is exposed along the river and on U.S. 2, northeast of the Great Eddy. The Sangerville formation is best exposed on the south bank of the river, just north of U.S. 201 and southwest of the Great Eddy. The stream eroded the gorge along the contact between the two formations.

Norridgewock Falls in Madison and Anson on U.S. 201A tumbles across northeast-trending ledges of the Sangerville formation. It was here that Benedict Arnold and his men realized that their heavy bateaux, which were made of green wood only a couple of weeks before the expedition left Pittston, were leaking. It took them seven days to portage around the falls, repair their bateaux, and recuperate from the ordeal.

Above Norridgewock Falls, the river meanders back and forth across a nearly flat floodplain. In Maine, such broad flats along rivers are called intervales. The bedrock obstruction of the falls at Anson and Madison slows erosion of the channel and traps sediment to make the intervale. The river wanders about on its extremely flat surface in a series of looping meander bends.

Sands of Avalon

The Carrabassett River joins the Kennebec River along this sinuous stretch. You can see North Anson Gorge from the west side of the bridge on U.S. 201A, where it crosses the Carrabassett River. The rather broad and shallow gorge cuts into vertical beds of the Madrid formation. As you look down on the outcrop, the sandstone layers appear as broad stripes of pale gray; the thin slate layers appear as thin stripes of dark gray.

Geologists who have studied these rocks and others like them have concluded that turbidity currents flowing northwest deposited the sand. Older sedimentary formations in this region were deposited from currents flowing in the opposite direction. The Madrid formation is the first package of sediment to come from the Avalon terrane.

From Maine 16, you can walk to the western edge of the gorge, just east of the railroad trestle. There the sandstones are in angular blocks suspended in a matrix of slate. The rock is a mélange, a chaotic mixture of rocks. You can see another spectacular mélange in the Madrid formation beneath the bridge over Gilman Stream in North New Portland, also on Maine 16. Both mélange zones are more than 100 feet wide, and quite impressive. The magma for the dikes and

Sandstone and thin slate beds of the Madrid formation exposed on the Carrabassett River in North Anson.

sills of igneous rock that cut through the mélange in North New Portland probably came from the Lexington pluton to the west.

Arnold's Landing

On October 9 or 10, 1775, Benedict Arnold landed along the ledges at the base of Caratunk Falls to prepare for his last portage on the Kennebec River. You can easily reach those ledges from a short road that heads west from U.S. 201 to the power plant.

The Madrid formation is nicely exposed on the lowermost ledges, where sedimentary features such as ripples, graded beds, and large crossbeds are beautifully preserved and displayed. The beds dip toward the river but get younger away from the river, which means that they are on the overturned limb of a fold. As you walk toward the woods and away from the river, you will find massive beds of slate. They are in the Carrabassett formation that overlies the Madrid sandstone.

While working on these rocks a few years ago, a geologist encountered "B. A. 1775" deeply carved into an obscure portion of the ledge. Could it be?

Seawater in the Kennebec River

About 12,500 years ago, as the ice sheet melted back through this part of Maine, seawater followed the retreat inland. Glacial marine mud occurs at least as far inland as Bingham, perhaps to Caratunk. During low water, you can see the gray clay in the channel of Austin Stream, which enters the Kennebec River at Bingham. These clay and silt deposits were a problem when the Wyman Dam was built.

A Weighty Departure

At the Great Carrying Place, 8 miles north of Bingham, on October 11, 1775, Benedict Arnold's party left the Kennebec River and headed west across the divide. They carried their supplies, already soaked and spoiling, and waterlogged bateaux that weighed more than 400 pounds each, over nearly 8 miles of rugged terrain to the Dead River. After reaching the Dead River, Arnold made camp and erected a flagpole. The township of Flagstaff was named for this campsite. The dam at the head of Long Falls flooded it in 1950, when Flagstaff Lake filled. Arnold and his men then traveled up the Dead River and over the rugged divide into the Chaudière River. Maine 27 from Stratton to Coburn Gore follows their path up the Dead River.

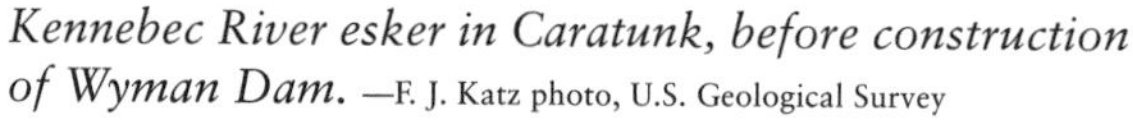

Kennebec River esker in Caratunk, before construction of Wyman Dam. —F. J. Katz photo, U.S. Geological Survey

Kennebec River Esker

The Kennebec River closely follows a large esker. In places, it makes islands in the river, as at Wyman Lake, between Bingham and Caratunk. South of Bingham, the esker is on the west side of the river, where it looms over Maine 16A. North of Bingham it is on the east side of the river, under the road in a few places. It continues north to Brassua Lake. The esker system extends south to end near Merrymeeting Bay, north of Brunswick.

Grungy Rusty Rock

You can see the Great Carrying Place, just north of Carrying Place Stream, from the precipice that U.S. 201 follows along the east side of Wyman Lake. When Wyman Dam impounded the Kennebec River in the early 1930s, it forced the road out of the valley and onto the canyon walls that surround it. Although the Madrid and Carrabassett

Cliffs of the Smalls Falls formation near Wyman Lake, Moscow.

formations are exposed between Moscow and Caratunk, the rocks best exposed along this section are the mudstones of the Smalls Falls formation. The dark and rusty ledges that overlook the Great Carrying Place are unmistakably the Smalls Falls formation; as in most exposures, the closer you get, the uglier it looks. A depressing rusty stain and sulfidic grunge obscures all the details of the rock.

Weathering of the iron sulfide mineral pyrrhotite releases sulfuric acid. About the only thing that can tolerate these harsh conditions is the white birch. In Concord Township, near the west abutment of the Maine 16 bridge over the Kennebec, there are other good outcrops of the Smalls Falls formation, some with faults cutting across the bedding.

The Smalls Falls formation was the last of the western sequence to be deposited before the Madrid formation. Loaded with unrecognizable organic matter and rich in iron sulfides, the formation was deposited in a marine basin in which the water contained very little oxygen. Why, no one knows. By the time the Madrid formation was deposited, oxygen was more available, again for reasons no one understands.

The Forks Formation

A striking black and silvery turbidite is exposed in a narrow band that passes through The Forks and gives this formation its name. There is an excellent exposure of this rock on the east side of U.S. 201, about 300 yards south of the bridge over the Kennebec. The town is named for the juncture of the Kennebec and the Dead River that enters from the west.

Kennebec River Raft Trip

A number of rafting companies run day trips down the Kennebec River, a wet journey of about 10 miles. The trip begins at Harris Dam, an electrical generating facility. The river eroded this gorge in the Dead River formation; the vertical bedding and fractures in the rock control the steep walls. The river makes nearly 90-degree turns in the gorge where it eroded along one set of fractures or another that cross at right angles. The depth of the gorge limits scenic views.

Moxie Falls are within a half-mile of the gorge; they are about 90 feet high, one of the highest in Maine. The water tumbles across The Forks formation, a sequence of interbedded sandstones and muds that were deposited during Silurian time, about 420 million years ago. The beds are vertical, and they contain a number of pale dikes that formed as magma intruded fractures during Devonian time, about 380 million years ago.

Dark and light beds in The Forks formation near type locality. The small drill hole was made to find where on earth the sediments were deposited, from a study of the magnetism in the rock.

Kennebec rafting trips usually end at The Forks, the juncture of the Kennebec River with the Dead River, which comes in from the west. Rafting and white-water canoe races also take place on a wild stretch of the Dead River where the water certainly belies the image of the name. In late summer and fall, the flow is generally too low for river running without water released from Flagstaff Lake.

Lobster Mountain Anticline

North of The Forks, U.S. 201 climbs steeply to a plateau where the bedrock is sedimentary formations deposited during Cambrian and Ordovician time. They are exposed in the Lobster Mountain anticline, actually quite a belt of folds. The rocks that outcrop along the road include the Dead River and Hurricane Mountain formations. The Dead River formation is greenish turbidite mudstones; the Hurricane Mountain formation is a tectonic mélange, a mess of rocks that were swept off the sinking floor of the Iapetus Ocean and stuffed into the trench. Nearby rocks include oceanic basalts, a scrap of the floor of the Iapetus Ocean. They are called the Jim Pond formation. The same formation is well exposed along Maine 27 north of Eustis.

Johnson and Coburn Mountains and the Tarratine Formation

Some of the youngest sedimentary rocks in this part of Maine appear along U.S. 201, within the Moose River syncline, another fold belt. The most conspicuous is the Tarratine formation, a sandstone with thin layers of slate deposited during early Devonian time. This resistant formation holds up Johnson and Coburn Mountains, southwest of the road. The rocks are folded into a series of anticlines and synclines that continue northeast to Moosehead Lake. The road crosses the Tarratine formation from just north of West Forks to the top of the long grade south of Parlin Pond, past the pond to the top of the next long grade that reaches the height of land that overlooks Attean Pond and the Moose River.

Many outcrops of the Tarratine formation have fossil brachiopods, in particular near the top of each of the sandy beds. Glaciers eroded blocks of the Tarratine sandstone with fossils and distributed them far and wide over the Central Maine slate belt. Many gardeners have found specimens of this "fossil rock" when working the tilly soil.

Fossils in the Hardwood Mountain and Beck Pond Formations

Two fossiliferous limestones appear about 15 miles west of Parlin Pond, west of Spencer Lake. Spencer Lake is beautiful and remote, worth the trip. The fossils are an added attraction.

The Hardwood Mountain formation contains fossils of corals that lived during Silurian time. The original sediments were deposited on the Attean pluton, which intruded the older rocks of the area as a mass of granite magma during Ordovician time. Obviously, the Attean pluton, which crystallized at some depth, had lost its cover to erosion by Silurian time.

The Beck Pond limestone was deposited during Devonian time. It also contains fossil corals, as well as fragments of granite that came from the Attean pluton. Perhaps the Beck Pond limestone was laid down around islands of the granite. Imagine coral reefs growing in a warm and shallow sea between little islands of granite. Maine has not been the same since.

Attean Pluton

The Attean pluton is one of the few large bodies of granite in the state that date from Ordovician time, when the Taconic mountain-building event was underway. Attean is apparently an American corruption of *Etienne,* French for "Stephen."

Watch for the rock near the crest of the long grade from the south and in a broad area of low ground northwest of the highway. It is much fractured, with rust on the joint surfaces.

Pink and green granite of the Attean pluton, Jackman.

The Attean pluton was originally a quartz monzonite, a rock with plagioclase and orthoclase feldspar in about equal abundance. The plagioclase has been altered to quartz and yellowish green epidote and the orthoclase is pink, making a rock called unikite. Some people think it looks like salmon and peas, the old traditional Maine meal for the Fourth of July. The first sweet peas were ready then, and the salmon were running in the rivers. In recent years, peas are rarely ready for the Fourth. The climate is colder now than when the meal became the custom.

A long belt of Ordovician rocks a few miles south of the Attean pluton contains some volcanic rocks. They apparently erupted in a chain of volcanic islands, above a sinking slab of oceanic crust, during the Taconic mountain-building event. The Attean pluton probably crystallized from one of the masses of magma that rose into the crust during that episode.

Catheart Mountain Molybdenum

To the east of Bog Number 5 and almost due south of the Attean Pond overlook is Catheart Mountain (pronounced *cat*-heart). It is near the eastern margin of the Attean granite pluton. Like many granites, this one contains molybdenum, which is an essential ingredient in some alloy steels. The most abundant molybdenum mineral in this deposit is molybdite, the bright yellow oxide. The most common molybdenum mineral elsewhere is molybdenite, the dark silvery gray sulfide. It occurs in the Katahdin granite to the northeast.

Some geologists have wondered whether a body of molybdenum ore may lie beneath Bog Number 5, near the center of the Attean pluton. Assays show that the peat there contains far more than the normal amount of metals. Evidently, groundwater circulating through the granite carried dissolved metals into the peat, where the organic matter absorbed them. Environmental and economic constraints would probably prevent mining, even if good ore does exist.

Moose River

The Moose River is the largest headwaters tributary of the Kennebec River. It flows from the area about 20 miles west of Jackman, on the east side of the Boundary Mountains. Moose River flows through a number of lakes before reaching Moosehead Lake and the beginning of the Kennebec River. Below Moosehead Lake, the Kennebec River flows all the way to the sea without passing through any natural lakes. The Penobscot and the Androscoggin Rivers flow through a number of lakes in their headwaters, but none after they leave the mountains. The portions of these rivers without lakes corresponds to the area of marine submergence. It is possible that some potential lake basins here were filled with glacial marine mud. Large stranded ice blocks prevent sediment from filling potential basins. If these blocks were floated by the deep marine waters, the depressions could then fill with sediment.

The Switzerland of Maine

The town of Jackman is known as the Switzerland of Maine because it is surrounded by mountains and lakes. A large portion of the population speaks both French and English. Log truck drivers from Québec and America feel at home.

Attean and Wood Ponds are two of the largest in Maine without dams. The Moose River flows unhindered through Jackman and its ponds and then on east to the Kennebec.

Frontenac Formation

Bedrock at the border crossing is slates and phyllites of the Frontenac formation, a turbidite mudstone deposited during Silurian time. Outcrops on the Québec side display obvious folds. Many of the folds have smaller folds riding piggyback on them.

Overlying the bedrock is the bouldery till of a moraine deposited from the last ice sheet near the end of the last ice age. As the ice melted back to the north, down the tributaries to the St. Lawrence River, the ice front acted as a dam that blocked meltwater, forming temporary glacial lakes. North of St. George, drainage of the Chaudière River was blocked and a large delta was deposited in the lake. It is now a ridge that extends nearly across the valley.

Maine 15
Abbot Village—Jackman

75 miles

Nearly all the bedrock between Abbot Village and Greenville consists of the black slates of the Carrabassett formation. The Madrid formation of Silurian age underlies small areas.

Glaciated Outcrops

Glaciers eroded some exposures of the Carrabassett formation just north of Abbot Village into excellent stoss and lee topography. The rounded stoss side faces north, in the direction from which the ice came. Particles of sediment embedded in the passing ice polished that surface and cut grooves and striations in it. The ragged south sides face south, in the direction of ice flow. Ice froze fast to the bedrock, then plucked fractured blocks as it flowed along. You can infer the approximate direction of glacial flow from the shapes of these outcrops, the exact direction from the compass orientation of the striations on their stoss slopes. Many larger rock hills in the area show the same glacial shaping on a much larger scale. You can find north by looking at their shapes.

Moosehorn

Moosehorn is the name of a fork in the road, marked in 1818 by a set of mounted horns pointing the way to Monson. Henry David Thoreau, on his first trip to the West Branch in 1853, passed this way and noted: "At a fork in the road between Abbot and Monson . . . I saw a guidepost surmounted by a pair of moosehorns, spreading four or five feet, with the word 'Monson' painted on one blade and the name of some other town on the other." In fact, on the other horn

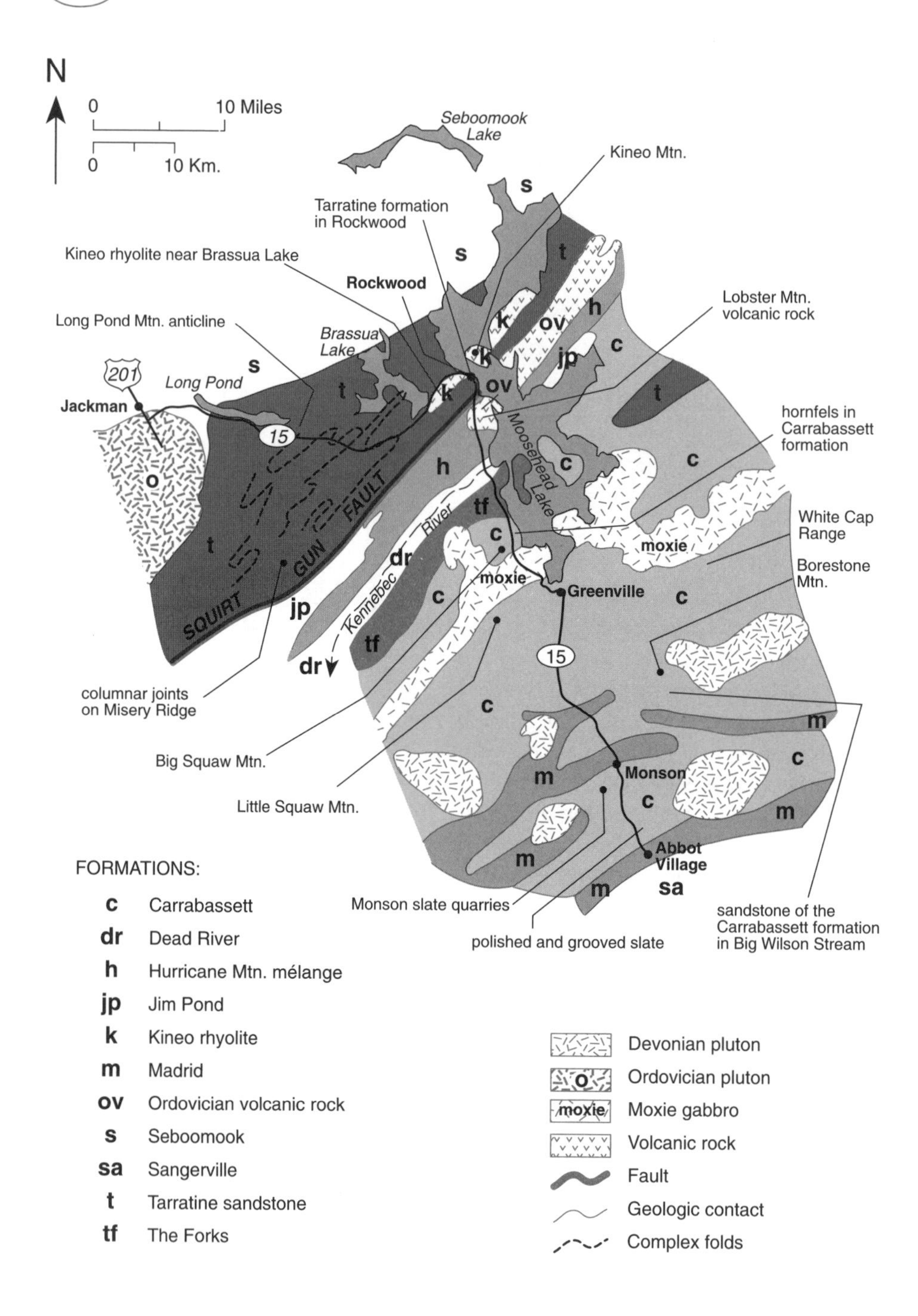

Geologic features along Maine 15 between Abbot Village and Jackman.

Glaciated outcrop of Carrabassett formation north of Abbot Village.

was written "Million Acres," which later became the town of Blanchard. The horns are stolen periodically, but always replaced, although the names have not appeared for some time.

The road from Moosehorn has been abandoned, so Blanchard now can be reached from Monson and Abbot Village. The Abbot road crosses the Piscataquis River at Barrows Falls, a beautiful little gorge with giant potholes carved into a section of the Carrabassett formation that consists entirely of broken and disrupted layers.

Slate Quarries of the Carrabassett Formation

Monson, on the Moosehead Trail, and Blanchard to the west were the sites of a remarkable industry that mined slate for electrical switchboards and pool tables. These were the only quarries in the country that produced slate of such high quality; the only others were in France.

Moses Greenleaf, an early entrepreneur devoutly interested in developing the resources of interior Maine, found the slate in 1814. His attempts to develop a slate industry in Brownville failed because both money and skilled labor were in short supply. When word of the discovery reached the slate districts of northern Wales, the skilled workmen there began immigrating to the Brownville area in the 1840s. They made the slate industry blossom. Three quarries opened. Much later, in 1872, a Welshman named Griffith Jones found excellent slate

A moosehorn sign in Abbot Village has stood here for at least 150 years.

in Monson, and the first of eighteen quarries opened in that district. From that time until the late 1920s, quarries operated along a narrow belt that extends southwest from Lakeview Plantation, through Brownville and Monson, to Blanchard.

It is now known that this belt of quarries follows a unique horizon within the lower Devonian Carrabassett formation. It was originally deposited as thick beds of mud that were later metamorphosed and gently folded during the Acadian mountain-building event. The slate occurs in a zone about 100 to 200 feet wide, on the flank of a synclinal trough folded in the rocks. The best slate came from a seam only 8 feet wide; it is so uniform that nuts and bolts can be machined from it.

Although slate is one of the most abundant rocks in Maine, little is good enough to mine. Most contains thin layers of sandstone or limestone. Or it breaks unpredictably into irregular shapes. Good slate must be uniform in composition and color, and it must split into thin slabs.

Roofing shingle was only the by-product of these quarries. They mainly produced slate for billiard tables, blackboards, telephone switchboards, floor tiling, sinks, refrigerator linings, signs, memorials, foundation pieces, and many other uses.

Although most of the roofing shingle was shipped to Boston, it was also used locally. Watch for it on nearly every house roof in the Monson and Brownville areas. After seeing the elegant black Monson slate in the John F. Kennedy Memorial in Arlington National Cemetery, I am amazed at the scarcity of slate memorials in local cemeteries. Most gravestones of the slate mining period are marble, which dissolves in the rain and was imported at much greater expense. Perhaps people thought the white marble more cheerfully heralded a brighter eternity than did the black slate.

Abandoned quarries in Maine eventually evolve into dumps or swimming holes. Until recently the town dump in Monson was the dry pit of the Kineo quarry, where you could enjoy driving to the edge and throwing your trash 150 feet down a vertical wall of slate. The trash accumulated as small deposits along the side of the pit floor; even after years of use the pile seemed minuscule compared with the size of the hole.

Black slate like that in the Carrabassett formation owes its color to carbon and finely disseminated iron sulfide. That is typical of sediments deposited beneath water that contains little or no oxygen. Red shales or slates, which are relatively rare in Maine, form from mud deposited in water that contains oxygen. Good examples of red sediment include the Perry formation in coastal Maine and Mesozoic rift basin deposits in the Connecticut River valley.

Turbidite Mudstone at Big Wilson Stream

Wilson Stream is about 8 miles northeast of Monson, on the Elliotsville Road. Excellent exposures of mudstone are beneath the bridge on the Elliotsville Road and at Little Wilson Stream Falls and gorge on the Appalachian Trail. Bedding and slaty cleavage are both vertical. Look for the excellent cross sections of graded beds and sand ripples in the water-worn bedrock pavements. Where the bedding planes are exposed on vertical surfaces, you can see ridges and grooves called flute casts and tool marks. Currents erode the flowing forms of the flute casts; pieces of passing debris carve the tool marks into the soft mud. If you look closely at the bedding surfaces, you can also see quite a variety of tracks and burrows, called trace fossils. These hard rocks were soft mud when they were deposited during Devonian time.

Borestone Mountain and the Onawa Pluton

Northeast of Monson lies a sharp, isolated peak called Borestone. Most published maps spell it Boarstone, but local residents insist upon Borestone. By whatever spelling, Borestone Mountain is part of a prominent rim of hornfels mountains that encircle an elongate basin eroded on a diorite intrusion called the Onawa pluton. The surrounding mountains are all eroded in rocks of the Carrabassett formation

Layers of sandstone at Big Wilson Stream, Elliotsville Township.

that the diorite magma invaded and then baked into hornfels. They stand high because the hornfels resists weathering and erosion. They include Borestone, Barren, and Benson Mountains, the first two of which are readily accessible by hiking trails.

Rocks on the precipitous north and northeastern flank of Borestone Mountain began their careers as deposits of mud. Then they baked and partially melted in contact with the hot magma of the Onawa pluton—diorite magma is much hotter than granite magma. The result is a rock that is partly metamorphic and partly igneous—a migmatite. You can see it nicely exposed along the crest of Borestone Mountain. As you move away from the contact of the Onawa diorite, you cross progressively less metamorphosed rock.

Rocks near the base of Borestone Mountain are hornfels in which you can see just a hint of slaty cleavage; heat from the Onawa diorite baked most of it out. As you climb the mountain, notice that the mineral grains in the rock become progressively larger, and the slaty cleavage disappears. Then little laths of andalusite, an aluminum silicate mineral, appear. At the top of the mountain, you can see little stringers of granite cutting through the hornfels. They crystallized from the molten magma.

The steep northeast slope of Borestone Mountain closely follows the contact between the pluton and its enclosing country rock. The

Thick sandy and thin slaty beds in Carrabassett formation on Big Wilson Stream, Elliotsville Township. Thin dark lines are ripples. The dark blotches are prints of a big wet dog.

diorite is now so deeply eroded that the only indications of it you see on the peak are the hornfels and stringers of granite, evidence of its heat.

Submarine Landslides and Debris Flows

The Carrabassett formation also contains layers of sandstone and mudstone that were chaotically mixed while still soft sediment. They look like they were stirred. These are submarine landslide deposits. Soft sediment slid down the continental slope and accumulated in a jumble on the deep ocean floor at its base.

The abundance of such deposits in the Carrabassett formation, and its status as the youngest and most widespread formation in the Central Maine slate belt, suggest that the Acadian mountain-building event was truly underway when it was laid down. Mountains were rising to the southeast. Rivers carried sediment from them to the coast, where it was deposited on submarine slopes. Then it moved down to the deep ocean floor, either in submarine slides or muddy turbidity currents.

Slate South of Greenville

Roadcuts on Maine 6 and 15 south of Greenville expose slump folds and debris flows in the Carrabassett formation. On Indian Hill, south of the Greenville rest area, you see dark gray slates with strong

Carrabassett formation dominated by slaty cleavage in Greenville near Indian Hill. Slaty cleavage nearly obscures the bedding planes.

cleavage that look rather featureless at 55 miles per hour. But a closer look, especially at the exposure west of the road, reveals that the folds have no apparent relationship to the slaty cleavage. Slaty cleavage is normally in the direction parallel to a plane that would split the folds into symmetrical halves. Evidently, these folds existed before the rock was metamorphosed into slate. They probably formed while the original muds were still soft. All the exposures of Carrabassett formation along Maine 15 west of Greenville Junction, along the north ends of Little and Big Squaw Mountains, show similar signs of soft sediment deformation.

Hornfels Mountains of the White Cap Range

Watch for the rest area at the crest of the hill where you first glimpse Moosehead Lake. The picnic area provides a spectacular view to the east of the hornfels mountains in the White Cap Range, nearly all of which are composed of the Carrabassett formation. The outcrops nearby are slate, which bakes into spotted hornfels only within a mile or so of the molten magma of a large igneous intrusion. In the Greenville area and in the mountains to the east, the heat source was the Moxie gabbro, a mass of gabbro that had a temperature of close to 2,000 degrees Fahrenheit when it was molten.

Greenville and Moosehead Lake

Greenville has long been the last outpost for loggers, fishermen, and hunters. Years ago you reached the woods by canoe, later by stagecoach, then up the long lake by steamboat. Train service began in the late 1800s. Float-plane service was available by the 1930s. Greenville and Greenville Junction are said to have the largest concentration of float planes in the country. Roads built in the last thirty years now reach nearly all the north woods sites.

Indians traveling with Henry David Thoreau in 1853 tried to convince him that Moosehead Lake got its name from the resemblance of Kineo Mountain to the head of a cow moose. It seems more likely that it was named for the resemblance of its map outline to the antlers of a bull moose. It is the largest lake in New England, about 40 miles long and with 350 miles of shoreline. Two dams on the west side control the level of the lake and release water into the Kennebec River. The East Outlet was the natural drainage from the lake, but a 15-foot dam raised the level until it also drained through the West Outlet, which then needed a dam of its own.

In the 20 miles between Greenville and Rockwood, the highway follows an old woods road that continues north to Canada. About 20 miles north of Rockwood this old road passes Pittston Farm, which was a prisoner-of-war camp during World War II; the German prisoners worked as loggers. The old woods road from Rockwood enters Québec near Dole, another woods camp. The area around Dole Pond is now one of the largest maple syrup producers in New England.

In the first 10 miles north of Greenville, Maine 15 crosses slates of the Carrabassett formation, baked hornfels near the Moxie gabbro, and the Moxie gabbro. On either side of the two outlets of the Kennebec River, erosion of the long and narrow Lobster Mountain anticline has exposed Ordovician and Cambrian rocks. These older formations are at the surface because they were raised in the core of the anticlinal arch. Near Rockwood, Maine 15 passes into Devonian mudstones and volcanic rocks that are younger than the Carrabassett formation. It continues across these rocks to Jackman.

Little Squaw Mountain

At Greenville Junction, the former Canadian National Railroad met the Maine Central Railroad. The Canadian railroad was built across Maine in 1889. Before World War II, the Maine Central line passed beneath the Canadian trestle and ended on the dock; from there, tourists and sportsmen boarded steamboats that took them to one of the hotels and sporting camps on the lake.

If you look west from the junction, you can see a prominent ridge of hornfels, Little Squaw Mountain. This is part of the Carrabassett formation, baked in the heat of the molten gabbro of the nearby Moxie gabbro. These hornfels ridges typically have bumpy crests. The rock at Greenville Junction is slate of the Carrabassett formation, but no outcrops are visible.

About a mile north of Greenville Junction, the road curves sharply around the northeast end of the hornfels ridge of Little Squaw Mountain. The outcrops along this curve show no slaty cleavage, because it was annealed out of the rock as it baked into hornfels. Bedding is much more obvious in the hornfels than in the slate outcrops of the same formation about 3 miles south of Greenville. The rock contains small crystals of light pink andalusite, a mineral that typically forms in slate subjected to high temperatures at moderate pressure.

Moxie Gabbro

The Moxie gabbro consists mostly of gabbro, the coarsely crystalline equivalent of basalt, along with a variety of other rock types. The magma almost certainly melted in the mantle, above the slab of ocean floor that was sinking into the mantle during Devonian time. Outcrops of the Moxie gabbro extend from Moxie Pond northeast almost to the Katahdin granite, 20 miles east of Greenville Junction. The intrusion is long and thin, locally more than 1 or 2 miles across, generally much less. A geologist I know calls it the Noodle.

The composition of the rock in the Moxie gabbro changes conspicuously along the length of the intrusion, from granodiorite on the northeast to dunite near Moxie Pond at the southwest end of the intrusion. As the name suggests, granodiorite is much like ordinary granite, except in being a bit darker. Dunite is a rare rock that consists mostly of the green mineral olivine. How is it possible for such radically different and normally incompatible rocks to exist within the same igneous intrusion?

One explanation proposes a scenario in two acts. First, while the original basalt magma was still extremely hot, the first minerals to crystallize, olivine and pyroxene, sank to the bottom of the magma chamber to make the dunite. That took much of the iron and magnesium from the remaining liquid magma, which then crystallized into granodiorite. At this point, the intrusion was dunite at the base and graded up into granodiorite. The second step tilted the crust up toward the southwest, raising the dunite. Then erosion planed the tilted intrusion, exposing dunite at the southwest end and granodiorite at the northeast end. If you travel from northeast to southwest, you are also traveling down through the continental crust.

Other evidence confirms that the rocks near the southwest end of the Moxie gabbro have been more deeply buried than those at the northeast end. For example, the formation that is slate in Greenville has been metamorphosed to sillimanite schist near the New Hampshire border.

An outcrop on Maine 15 exposes finely crystalline rock near the edge of the Moxie gabbro. It was chilled so quickly against the cool country rocks that the crystals could not grow big enough to be visible without a lens. This rock is a good representative of the original magma before any of the heavy iron and magnesium olivine crystals could settle. The outcrop is much fractured, weathering to a rusty soil.

You can see an exposure of the central part of the Moxie gabbro near the line between Little and Big Squaw Townships. The rock is composed of crystals large enough that you can easily see them without using a magnifying lens. It is gabbro, nearly intermediate in composition between the granodiorite to the northeast and the dunite to the southwest. Weathering of this outcrop has accentuated some round structures in the rock that are knots of small crystals of olivine, pyroxene, and feldspar.

Above these is a thick brown soil formed from the complete weathering of the Moxie gabbro. The 12,000 years since the ice melted hardly seems time enough to weather such a thick soil. I believe the

Weathered outcrop of coarse-grained gabbro near the line between Little and Big Squaw Townships.

soil dates from before the ice ages, but I cannot explain why the glaciers did not bulldoze it off the landscape. Perhaps the soil was frozen when the ice went over it, or maybe the soil was much thicker and this is all that remains.

About halfway up the small hill on the other side of the valley formed by the weathering of the Moxie gabbro is another outcrop of the chilled border phase of the gabbro.

Squaw Mountain

Squaw Mountain, at 3,196 feet, is the highest elevation near Moosehead Lake. It is also the second highest elevation in the range of hornfels mountains that extends northeast to the Oakfield Hills near Houlton. Mountains of similar origin increase in height toward the southwest, culminating in Mt. Washington in the White Mountains of New Hampshire. The highest hornfels mountain in Maine, the second highest in the state, is Sugarloaf, at 4,237 feet.

The hornfels near the summit of Squaw Mountain was heated almost to melting; the layers of rocks lost their strength and formed flowing swirls near the molten basalt of the Moxie gabbro. Some of the chilled border phase of the Moxie gabbro, finely crystalline basalt instead of coarsely crystalline gabbro, is exposed near the summit. Evidently, the basalt resists weathering and erosion more successfully than the gabbro in the interior of the pluton, which tends to erode into lowlands.

Squaw Mountain owes its name to an Indian legend about a brave named Kineo, who left his tribe in a huff to live on the mountain that now bears his name. One night he saw a light at the south end of Moosehead Lake. On investigating, he found it was his dying mother's campfire. He buried her nearby, giving Squaw Mountain its name. Then he returned to his tribe, as his mother had asked.

Carrabassett Formation on the North Side of the Moxie Gabbro

Watch near the entrance to Big Squaw Ski Resort for outcrops of nicely bedded mudstones, now baked into a tough hornfels. The layers dip gently down to the southeast, toward Little Squaw Mountain, which you can see a mile or so away, on the far side of the Moxie gabbro.

The outcrop on the east side of the road is natural, and has not been blasted; numerous glacial striations are preserved on the steep face of the rock parallel to the road. The sedimentary layers are 1 to 2 inches thick. The sandy lower part of each one stands out as a pale band. If you look closely, especially at the north end of the outcrop,

Hornfels of Carrabassett formation in Big Squaw Township.

you can see that the lighter, sandy lower part grades upward into darker, finer mud, the typical pattern in mudstones deposited from turbidity flows.

Blocky little crystals of pink andalusite, an aluminum silicate mineral, stand out in the muddy upper parts of the layers. They formed there because the mud contained clay, which contains aluminum. It is easy to be confused by these large andalusite grains. They make it appear that the coarsest sediment is at the top of each turbidite layer. In the original sediment, the sandy bottom of each bed was coarser than the muddy top. Metamorphism did not change the size of the sand grains, but turned the mud into large crystals, giving the false impression that the beds are upside down.

The other minerals with the andalusite in the muddy layers include little flakes of black biotite, dark gray crystals of glassy quartz, and cordierite, which looks about like the quartz. On a sunny day, fresh pieces of this hornfels sparkle brightly, the sun reflected from the biotite flakes.

The broad folds in the layers apparently formed while this rock was still soft sediment. Layers above and below the folded layers are still straight. Evidently, the folded layers slid on the layers beneath and were then covered by new layers of sediment that did not slide.

Close-up of hornfels from Big Squaw Township. Light beds are sandy; darker ones are muddy. Light spots are pink andalusite crystals.

Hurricane Mountain Formation

Maine 15 crosses the East Outlet of the Kennebec River north of Greenville. The low dam and the trestle of the former Canadian railroad are just upstream from the highway. The rail line was built across Maine in 1889 as a shortcut between the Maritime Provinces and the western part of Canada. A tectonic mélange, the Hurricane Mountain formation underlies the outlet of the Kennebec River but is poorly exposed here. It was named for a mountain near Pierce Pond, between Flagstaff Lake and the Kennebec River.

The Hurricane Mountain formation consists of blocks of basalt, gabbro, sandstone, and other kinds of rocks set in a matrix of slate. It is a chaotic mess. These blocks were caught in an oceanic trench and mixed with the mud as the sinking slab of oceanic crust dragged them down. That happened during an event of Cambrian time called the Penobscot mountain-building event.

Islands in Moosehead Lake offshore from the East Outlet, and in Indian Pond a couple of miles downstream from the bridge, are made of the same mélange. The slaty matrix is so nonresistant that it erodes more quickly than the blocks. Green and Black Islands consist mostly of blocks of basalt and gabbro, respectively. They are in a cluster of other exotic block islands in Moosehead Lake, within a mile of the East Outlet.

The outcrops of the Hurricane Mountain formation are near the axis of the valley of the Kennebec River, as well as in the broad, long arm of Moosehead Lake called Spencer Bay. Both are in the axis of the Lobster Mountain anticline. This suggests that the rocks along the anticline are not very resistant to erosion. In fact nearly all of the rocks in this belt are quite weak; they underlie a broad, long lowland that extends 40 miles or so to the southwest and 30 or so miles northeast of the East Outlet. The slaty rocks along this trend were deformed at least twice during separate collisions. Water easily penetrates these rocks, allowing them to weather and weaken.

Lobster Mountain Volcanic Rocks

As you might suspect, the Lobster Mountain anticline was indeed named for Lobster Mountain. So were the Lobster Mountain volcanic rocks exposed in it. Geologists strictly observe their custom of naming rock formations and structures after geographic localities, lest some egotistical colleague name them for himself.

The Lobster Mountain volcanic rocks that you see along Maine 15 include greenish andesite and basalt that erupted along an island arc during Ordovician time, while the Taconic mountain-building event was in progress. They continue northwest in a narrow belt that reaches Lobster Mountain.

Lobster Mountain volcanic rocks in Taunton and Raynham Academy Grant.

The andesite was gray and the basalt black when they were fresh; now that they are slightly metamorphosed, the original black minerals, mostly pyroxene, are altered to green chlorite, a member of the mica family of minerals. Chlorite is one of the commonest causes of green colors in rocks. Some of the basalt has a sort of cellular structure that developed from gas bubbles that were trapped in the cooling lava. The gas was mostly steam.

These volcanic rocks are the most resistant of any in the Lobster Mountain anticline, forming low hills and mountains. Moosehead Lake narrows considerably where it crosses the Lobster Mountain volcanics.

Tarratine and Matagamon Sandstones

The Tarratine formation is a sandstone named for a railroad siding near Brassua Lake. The sand was deposited during Devonian time, probably as a delta in shallow water. Unlike the mudstone of the Carrabassett and Seboomook formations that underlie it, the Tarratine sandstone contains fossils, in abundance. They include impressions of brachiopods, snails, and other creatures that lived in the warm, shallow sea that lapped onto the delta surface.

A similar rock, also full of fossils, the Matagamon sandstone, appears to the northeast, around the north end of Baxter State Park. Both formations were once included in the Moose River sandstone, named after outcrops on the Moose River. It tends to form ridges, which made the formation an easy target for glaciers. They carried

Tarratine sandstone in Rockwood. Bedding is inclined steeply to the left.

Brachiopod fossils in Tarratine formation carried to Monson by glaciers.

pieces of these sandstones farther south, where you can now find them in the glacial till, as well as in streams, along lakeshores, and in stone walls and rock gardens.

The Moose River sandstone is probably the most widespread and well known of all the fossil rocks in Maine. Geologists working in the central and southern parts of the state are often asked to identify pieces of it.

The source of the sand in the Tarratine and Matagamon formations was probably the Avalon terrane, which was then approaching from the east during the Acadian mountain-building event. Sand ripples and other indicators show that currents were moving the sand from the southeast. Similar flow directions have been inferred for the deeper water sediments in the Carrabassett and Seboomook formations. This kind of information can be combined with knowledge of the folds and faults to tell something about the geologic history of the region.

Kineo Flint

The Devonian Kineo rhyolite directly overlies the Tarratine formation and is named for exposures on the towering cliffs on Kineo Island. Southeast of Kineo Island more rhyolite occurs on Blue Ridge, a 3-mile-long trend that culminates in high cliffs over Brassua Lake. Northeast of Kineo is a line of steep-sided hills that the observant Henry David Thoreau described as "a whole family of Kineo Mountains."

Kineo rhyolite worked into large point. This was found on the old shoreline of Brassua Lake.

Kineo rhyolite in Taunton and Raynham Academy Grant, near Brassua Lake. Light and dark bands are ash beds that accumulated after explosive eruptions.

Although the Kineo rhyolite erupted in violent explosions of steam and volcanic ash, the rock now has the mechanical properties of flint—a sedimentary rock. Most people call it the Kineo flint. Sharp blows strike rounded chips off the rock, leaving a thin edge. This property made the Kineo flint a favorite of ancient Indians, who traded it up and down the East Coast. A few years ago, Brassua Lake was lowered to its natural level while the dam was under repair. Archaeologists from the University of Maine found dozens of workshop sites along the old shoreline, where Indians had chipped the Kineo flint into implements.

Fresh pieces of Kineo flint are dark gray, many with lighter wavy bands. The surface of most pieces has been weathered to a chalky white. The Kineo flint is technically a tuff, or volcanic ash. In this case, the ash was still partially molten when it settled, so the particles welded together. Some cliff and roadcut exposures of the Kineo flint break into vertical columns in the manner of many lava flows. Many specimens contain visible grains of quartz and feldspar, as well as small blobs of granite that were torn from the walls of the magma chamber and then erupted with the volcanic ash. Outcrops of the Kineo flint show thin bands. Some of those are squashed pieces of pumice; others may be layers of older ash that were heated as the new ash covered them. The Kineo flint, more accurately the Kineo rhyolite, closely resembles the Traveler rhyolite north of Katahdin. It also occurs to the northeast on Big Spencer Mountain.

Moose River

Maine 15 follows the Moose River, the major source of Moosehead Lake. The river is lined with float-plane bases and fishing and hunting camps. A narrow bridge crosses Moose River near Rockwood and leads to Tomhegan Township and Pittston Farm; in the old days it was the route to Canada by the woods settlement of Dole.

Brassua Lake

Brassua Lake was originally two small lakes. Then the dam on the Moose River raised their level and combined them. Its shores have more rock outcrops than those of most Maine lakes. A high cliff of Kineo flint overlooks the east shore; quarrying rock for the dam made it higher. On the other shores are outcrops of the Tarratine sandstone and the overlying Tomhegan mudstone, which is also Devonian in age and the youngest rock in this part of the state. The Tomhegan mudstone is a mudstone with a finer texture than the Tarratine sandstone.

Misery Ridge, south of the road, is eroded in the Tarratine sandstone. This ridge, which continues for almost 50 miles, resembles those in the central Appalachians and formed essentially in the same way: the resistant sandstone was folded during the plate collision that caused the Acadian mountain-building event and now stands as a ridge because it resists erosion.

The streams that eroded some of the gaps in Misery Ridge now flow elsewhere. The Canadian railroad passes through one of those dry gaps. Other gaps still have streams flowing through them, such as Chase Stream. The pattern of streams near and on Misery Ridge is angular, with many sharp turns. Melting of the glacier at the end of the last ice age, about 12,000 years ago, flushed meltwater and outwash sediments through some of the gaps in Misery Ridge, perhaps contributing to their erosion. Several eskers and other meltwater deposits continue through the gaps and then turn to parallel the ridge.

A section of esker makes a line of small islands at a roadside campsite near the west end of Brassua Lake, along Maine 15. This esker continues along the north side of Misery Ridge to the vicinity of Misery Pond, where it becomes part of the Kennebec River esker system.

Long Pond Mountain

About 4 miles northwest of where Maine 15 departs from Misery Ridge at Misery Stream the road goes through a high roadcut in Long Pond Mountain, which is also made of the Tarratine sandstone. The beds on the south end of the outcrop dip south, toward Misery Ridge, whereas on Misery Ridge these rocks dip north. Thus the rocks between the two ridges have been folded into a syncline that is named for the Moose River that runs along the axis of its trough.

The outcrop at Long Pond Mountain is at the crest of an anticline, as you can see by looking for the layers of sandstone that tilt down to the south, toward Misery Ridge, and down to the north, toward the Moose River. More outcrops of Tarratine sandstone occur in the Moose River at the Long Pond dam. These also dip down to the north, under the youngest sedimentary rock in the area, the Tomhegan mudstone.

Chase Pond Sluice Road and Misery Ridge

About 2 miles northwest of the Long Pond Mountain roadcut is an intersection with roads leading north and south. These dirt roads are largely maintained by a paper company and are generally at least as passable as Maine 15. The road south, locally known as the Chase Stream Sluice Road, passes the more interesting rocks; the road north over the Moose River crosses large areas of glacial outwash.

About 2 miles south of Maine 15 is a sharp bend along a high roadcut. The rock is our old friend, the Tarratine sandstone, here with abundant fossils. The ridge is part of Long Pond Mountain, with the rocks dipping down to the south. The numerous fractures and small faults that break the rock reflect the brittleness of this sandstone.

About 5 miles farther south, Misery Ridge comes in sight, with Misery Pond occupying a gap in the ridge. A stop at the campsite on the south side of Misery Pond leads to an interesting climb. Walk around the shore of the pond to a steep slope covered with large boulders fallen from the cliffs. The boulders are a talus. Continue up the talus, heading toward the bedrock cliffs.

The talus contains boulders of sandstone from the Tarratine formation and rhyolite in large columns that fell from the outcrop at the top of the slope. The rhyolite is a dike that formed where magma

Columns of garnet rhyolite on Misery Ridge, above Misery Pond, in Misery Township.

filled a fracture in the Tarratine sandstone. The columns lie almost horizontally, with their ends pointing toward the cooling surface at the edge of the dike. The rhyolite resembles that of the Kineo flint to the east, except it contains garnets. The rock at the summit of Misery Ridge is the Tarratine sandstone.

Squirt Gun or Southeast Boundary Fault

The dirt road that meets the Chase Stream Sluice Road 2 miles from Misery Ridge is nearly straight for about 24 miles, following the trend of the Southeast Boundary fault. Some call it the Squirt Gun fault because a geologist found a plastic squirt gun nearby. The fault separates Devonian rocks on the north from pre-Silurian rocks to the south. Movement on the fault shoved the older rocks over the younger Devonian rocks.

Its movement also shattered the rocks along the fault, opening them to seeping water. The broken and weathered rocks were vulnerable to erosion, so streams eroded a straight valley along the fault zone. The road to the northeast returns to Maine 15 near the West Outlet of the Kennebec. Much of it is either on or next to eskers and other glacial meltwater deposits.

Hurricane Mountain Mélange

Cross the road that follows the fault and continue south on the Chase Stream Sluice Road about a mile, then turn left over a wooden bridge. Just before you reach the top of a ridge that is almost completely clear-cut, you will see outcrops of the Hurricane Mountain mélange. Look for small exposures of slaty rock that is the matrix of the mélange. Small knobs in the clear-cut areas are blocks of exotic rocks such as basalt, gabbro, and sandstone.

From this ridge you see a spectacular view of the mountains on either side of the Lobster Mountain anticline. To the south, 20 miles away, is the ridge of Carrabassett formation that was baked into hornfels in the heat of the Moxie gabbro. The highest peaks in this trend are on Squaw Mountain, which has several separate peaks.

The small bumps along the trend are characteristic of most hornfels ridges. They form where the rock was more thoroughly hardened during contact metamorphism. To the north, across the Southeast Boundary fault, is Misery Ridge and others eroded in the Tarratine formation. The topography in this belt, from the mountains on the north, across the broad valley eroded in older rocks, to the hornfels mountains on the south, greatly resembles the Valley and Ridge Province of the Appalachians, on a somewhat smaller scale.

Dead River Formation

After the road descends from the Hurricane Mountain outcrops, it meets a somewhat better road. Turn left. Outcrops along this section of the road expose the Dead River formation—mudstones deposited during Ordovician time. The rock contains enough of the mineral chlorite to give it a distinctly greenish color. And this rock had bad experiences during both the Taconic and Acadian mountain-building events; each left it with a slaty cleavage. The two cleavages intersect, so the rock tends to break into long sticks, instead of into slabs like most slates. More outcrops of the Dead River formation exist at the end of the road, where it makes a loop at the Chase Stream sluice.

Chase Stream sluice was used during the log drives down the Kennebec River. Truckloads of logs were dumped on a broad metal slide that shot them into the air, then down more than 200 feet into the gorge of the Kennebec River below.

The gorge of the Kennebec River is about 10 miles long, and as much as 240 feet deep. It was eroded mostly in the Dead River formation, along the trend of its bedding and slaty cleavage. The present flow in the Kennebec River is probably large enough to account for the size of this gorge, since the rock here is not particularly strong. Nevertheless, glacial meltwater and the sediment it carried probably helped. Loggers helped, too. They blasted any rock that stuck into the channel to prevent logs from hanging up during a drive.

Mt. Katahdin, Baxter State Park, and Vicinity

Mt. Katahdin is the highest elevation in Maine; the top, known as Baxter Peak, is 5,268 feet and rises well over 4,500 feet above the surrounding lowlands. This is the greatest local relief of any mountain east of the Rocky Mountain Front. Charles Turner and his surveying party made the first recorded climb of Mt. Katahdin in 1804. Anyone making a first climb will appreciate why the Turner party estimated the height of the mountain as 13,000 feet. That group reached the summit by the Hunt Spur. In 1816 the great Abol Slide moved down the southwest side of the mountain, creating the favorite route up the mountain for many years to come, several miles shorter than the Hunt Trail.

Katahdin lies at the northern terminus of the 2,000-mile-long Appalachian Trail in Baxter State Park. The park is a wilderness of about

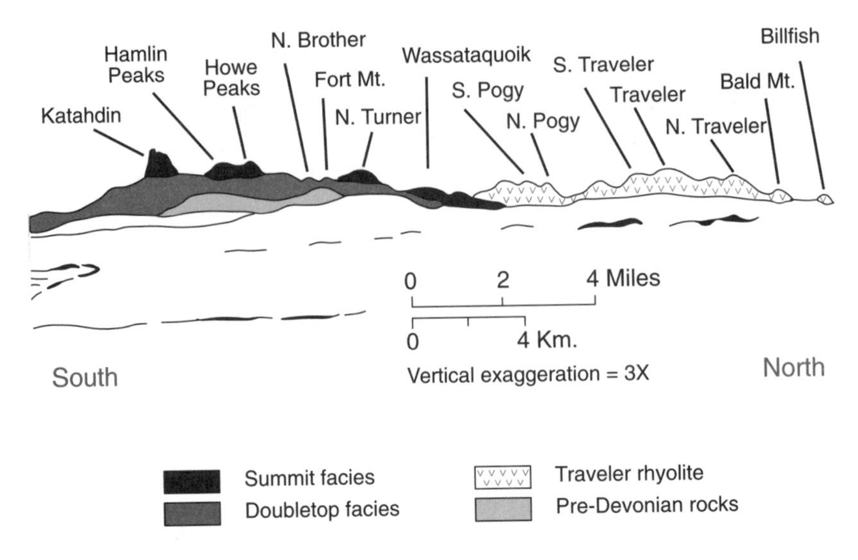

Cross section of Baxter State Park, from Katahdin in the south to Traveler in the north, a distance of about 22 miles.

200,000 acres, or 312 square miles. Percival Baxter, late governor of Maine, purchased it from paper companies and gave it to the people of Maine between 1930 and 1960. Paved roads nearly reach both the north and south entrances to the park. The perimeter road is little more than a narrow dirt track.

Devonian rocks underlie all of the park, except for a small area near the northern boundary. All rocks are associated with the Acadian plate collision, in one way or another. The Katahdin granite dominates the southern part of the park, including Mt. Katahdin; Devonian sedimentary and volcanic rocks dominate the northern half. Continental glaciers and small valley glaciers probably coexisted during the later portion of the last ice age, and both left their marks on the landscape.

Katahdin Granite and Mt. Katahdin

The Katahdin granite is exposed in an oval area about 40 miles long in a northeast-to-southwest direction, and about half as wide. The pluton extends east, west, and south of the park boundary, and far to the south of Mt. Katahdin and other high elevations that lie at its north end. Its cross section is shaped like a flat oval, with rounded lower and upper surfaces.

The molten magma intruded Devonian sedimentary rocks, as well as volcanic rocks that erupted from the same magma. The molten intrusion probably reached to within about 3 miles of the surface. The upper 500 feet or so of the intrusion cooled into glass, with no mineral grains at all. The interior crystallized slowly into a normal granite. All that happened about 400 million years ago.

Since then, erosion has removed the 3 miles of sedimentary and volcanic rocks that covered it, exposing the pluton. Meanwhile, the glassy top of the intrusion slowly crystallized into a granite with tightly interlocking mineral grains of medium size, the Summit facies. It supports the highest elevations on Mt. Katahdin.

Mineral grains deep within the pluton merely lie next to each other, without much interlocking. In places where erosion has stripped off the Summit facies, this weak interior rock—the Doubletop facies—has rapidly disintegrated. It is now eroded to a landscape of low hills, stream valleys, and lake basins. Where the Summit facies is still in place, as on the summit of Katahdin, it protects the Doubletop facies from erosion, much as some resistant rock caps the weak lower rocks in a mesa. The top of the pluton and the Summit facies slope gently down to the north, forming the Table Land, a broad upland surface on Katahdin.

Traveler Rhyolite

The Traveler rhyolite covers a broad area in the northern part of the park. It has the same chemical composition as the Katahdin pluton, is the same age, and apparently erupted from it.

The Traveler rhyolite erupted explosively, as flows of hot ash suspended in red-hot steam. Mount St. Helens produced similar eruptions in 1980. Exposures show layers of ash that contain collapsed lumps of pumice, fragments of quartz and feldspar that had begun to crystallize when erupted, and pieces of country rock ripped out of the crust. The ash was still hot, partially molten, when it settled, and these tiny pieces fused together on cooling. Outcrops of Traveler rhyolite tend to break into vertical columns like those you see in basalt lava flows.

The Traveler rhyolite fills a depression that may have been a collapsed caldera that opened as the magma beneath erupted. It has a total volume of about 80 cubic miles. Smaller bodies of similar rock exist northeast and southwest of the Traveler rhyolite. Together, they form the Piscataquis volcanic arc, named for the county in which it occurs.

Many geologists now think the Piscataquis volcanic arc erupted above a sinking oceanic plate that was plunging down to the southeast. Red-hot steam rising from the sinking plate melted rocks deep in the continental crust to form the Katahdin magma. The magma rose because it was lighter than the rest of the continental crust. Part of it erupted to become the Traveler rhyolite, then the magma below continued to rise and intruded its own volcanic cover.

Other rhyolites farther southwest have somewhat different chemical compositions, so they did not erupt from the Katahdin magma chamber. So far, no one has recognized their magma chambers, perhaps because they have been completely eroded, or possibly because erosion has not yet exposed them.

Matagamon Sandstone

Near the north entrance to Baxter State Park, you can see the initial flows of the Traveler rhyolite lying on top of the Matagamon sandstone. The layering in both formations is gently tilted, so they are only slightly deformed. Obviously, the Traveler rhyolite is on top of the Matagamon sandstone because it belongs there, not because the rocks are upside down.

Geologists interpret the Matagamon sandstone as a delta that was deposited from a landmass to the east. It covers the deep-water turbidite mudstones of the Seboomook formation. The Matagamon sandstone contains fossils that date it as early Devonian, and it must be older than the Traveler rhyolite, which lies on top of it.

The Matagamon sandstone underlies a broad area both east and west of the Traveler rhyolite. The similar and probably equivalent Tarratine sandstone to the southwest occupies a similar position beneath the Kineo volcanic rocks of the Piscataquis volcanic chain.

Trout Valley Formation

The Trout Valley formation is the youngest sedimentary rock in this part of Maine. It includes conglomerate, sandstone, and shale that appear to have been deposited on the floodplains of streams. Pebbles of volcanic rocks in the conglomerates suggest that the streams drained the mountains of the Piscataquis volcanic chain.

The lowest conglomerate rests directly on the Traveler rhyolite. It contains rounded pieces of rhyolite columns set in a matrix of ash. This conglomerate was probably deposited by debris flows, perhaps by debris flows that accompanied volcanic eruptions. The overlying sandstone and shales contain fossils of early land plants. Because both the underlying Matagamon sandstone and the overlying Trout Valley formation contain fossils, they bracket the Traveler rhyolite, and, in effect, the Acadian mountain-building event, with datable sedimentary rocks.

Ordovician and Silurian Rocks at Ripogenus Dam

In a number of places in this region, erosion of anticlinal structures has exposed rocks of Silurian and older age, in a sea of Devonian mudstones, volcanic rocks, and plutons of intrusive igneous rock. At Ripogenus Dam you can see rocks of Cambrian, Ordovician, and Silurian age exposed in the deeply eroded core of the Caribou Lake anticline. They tell a good story. Somewhat less accessible rocks formed during the same geologic periods are exposed in the Lunksoos-Weeksboro anticline along the Grand Lake Road east of the north end of Baxter State Park.

Ripogenus Dam is about 30 miles west of Millinocket, about 10 miles from Baxter State Park. Privately owned paved roads from Millinocket reach within 0.5 mile of the dam. Ripogenus Dam impounds a large reservoir on the West Branch Penobscot that was originally used for driving logs to lumber and paper mills downriver. Since about 1950, the dam has also generated electricity for the mills in Millinocket.

The oldest rocks exposed below the Ripogenus Dam are basalts that probably erupted from a rift at the crest of an oceanic ridge, or old oceanic crust. Geologists call them the Dry Way volcanic formation. The basalt flows contain widely scattered but well-formed pillows that leave no doubt they erupted underwater.

At a number of locations, the topmost basalt flows are weathered, so they were exposed to the atmosphere. The conglomerate that lies on the weathered surface is the base of the Ripogenus formation, which is early Silurian in age, about 430 million years old. The conglomerate consists mostly of quartz pebbles cemented with iron oxide similar to that in the underlying soil. A thin sandstone lies over the conglomerate, followed by a sandstone that contains lenses of highly fossiliferous limestone. The upper part of the Ripogenus formation is thinly bedded mudstones that contain some calcareous sand. Above them, andesite lava flows are exposed in the cliffs above Ripogenus Dam. These are called the Fat Man's Woe or the West Branch volcanic rocks. Their eruption was the beginning of the volcanism that accompanied the Acadian mountain-building event.

The complete section at Ripogenus, from Dry Way formation to Fat Man's Woe volcanic rocks, suggests great changes in the geography of Ordovician and Silurian time. The Dry Way formation was deposited in the deep ocean, probably at an oceanic rift, and was later exposed to weathering and erosion above sea level. The grain size of the rocks of the Ripogenus formation becomes progressively finer from bottom to top and suggests that the area was again submerged in deeper and deeper water. The Fat Man's Woe andesite contains pillow structures that indicate underwater eruption. The nearby Matagamon sandstone of early Devonian age was laid down in shallow seawater. All but the first few feet of the Traveler volcanic rocks was deposited from ash that fell on land, signaling that the area rose above sea level in Devonian time after having been submerged since late Precambrian time, a span of more than 200 million years.

The following scenario is proposed to explain the apparent changes in sea level in the Ripogenus Dam area. The crust was bent downward as the subduction zone under Avalonia approached. At first this flexure caused the crust in the Ripogenus area to rise just enough that the Dry Way volcanic rocks were raised above sea level, allowing the weathering found at their top. As Avalonia got even closer, however, it pushed the crust down, flooding the Dry Way and making room for the deposition of the Ripogenus formation. Sedimentation during early Devonian time locally filled the ocean, leaving just enough space below water for the first eruptions of the Traveler volcanic ash. The Katahdin area was not covered in seawater again until about 12,500 years ago, when the great ice sheet pushed the crust down again, allowing the sea to invade the East Branch briefly.

Glacial Features

C. T. Jackson reached the summit of Katahdin in September of 1835, during the first geological survey of Maine. He recognized that the mountain was composed of granite and also noted a number of boulders of other types of rock, mainly what is now recognized as the Matagamon sandstone. He interpreted these erratic boulders as sediment deposited from icebergs floating in Noah's Flood. It now seems abundantly clear that these erratic boulders were deposited from glacial ice that was thick enough to cover Katahdin.

A thick deposit of sandy and bouldery glacial till covers the lowland within and for some distance south of the Katahdin granite. The sand and boulders were weathered and broken mainly from the Katahdin granite. The sand also provided meltwater streams with the raw material for several systems of eskers, old glacial meltwater rivers, that terminate within Baxter State Park. They reach from the vicinity of Katahdin about 100 miles to the Pineo Ridge delta near Cherryfield.

Mountain glaciers erode deep basins in the heads of their valleys. They are called cirques. Unlike simple stream valleys, which are narrow and become smaller upstream, glaciated valleys are broad and end in a cirque, a wide amphitheater with nearly vertical walls. Small valley glaciers gouged about ten cirques on the higher slopes of Katahdin. The largest and freshest of these are on the east side of the mountain: North Basin, Great Basin, and South Basin. All three now receive large quantities of snow blown from the broad Table Land surface by the strong northwest winds. If the same winds blew during the last ice age, the valley glaciers in these cirques would have received more snow than those on the west and northwest sides of the mountain.

Landslides

Numerous landslides left scars on the steep south and east slopes of Katahdin and other mountains in Baxter State Park. The thin deposit of glacial till that covers these slopes slides down the mountain, exposing the pale bedrock underneath. The glacier smoothed the bedrock surface before plastering it with glacial till, so the till slips easily. Most slides happen in the spring, when the till is wet and heavy and perhaps has a heavy cover of wet snow.

Mysterious Patterned Forests

On the high slopes of Katahdin, especially around the large swampy Klondike Basin, you can see more or less horizontal bands of dead trees about 30 to 100 feet wide downhill of mature spruce trees of about the same age. Young trees growing in each band of dead trees increase in age and height downhill, until they blend into the mature forest below. It appears that the mature trees uphill of any dead tree band eventually die, joining the dead band, which is gradually replaced with new growth from below. The wave of dead trees coming up the mountain will eventually catch these new trees, continuing a strange uphill march of death and rebirth. Similar bands exist on other high mountains in New England.

Arctic Plants and Animals on the Table Land

Areas above about 3,000 feet on Katahdin and Traveler are above tree line. The shrubs, grasses, and dwarf trees that grow in this bleak environment are like those that grow in the arctic tundra, as are many of the insects and small rodents. How did these plants and animals get so far from home?

As the last great ice sheet began to form and move south about 30,000 years ago, it pushed an arctic environment ahead of itself. As that ice sheet melted back, after about 15,000 years ago, the arctic plants and animals colonized the newly exposed land. But the plants of the modern spruce, fir, and pine forests soon overwhelmed them. The tundra plants, and the animals that associate with them, survived only on the cold summits of high mountains. So the tundra plants and animals on Katahdin exist as though on an island.

Radial Stream Pattern on Katahdin

Maps showing the streams draining Mt. Katahdin are often used in textbooks to illustrate a radial stream pattern. Because the mountain is an isolated dome, water drains from it in all directions. Roaring Brook and Avalanche Stream flow eastward down the mountain slopes, Abol and Katahdin Streams flow southward, another Roaring Brook and Little Nesowadnehunk Stream flow westward, and Wassataquoik Stream flows to the north. The perfect example of radial drainage occurs on volcanic peaks like Mt. Rainier in the Cascade Range.

The Golden Road
Millinocket—Ripogenus Dam

30 miles

River transport of pulpwood to paper mills in Millinocket and East Millinocket ended in 1971. By that time a vast network of roads for hauling pulp had been woven through the forest. The Golden Road is the main road, so called to honor its cost. Meet the Golden Road at Millinocket Lake, pay the toll, and drive to Ripogenus Dam. The mile markers show the distance to the mill in Millinocket.

The road from Millinocket to Ripogenus Dam begins in slates that were deposited as mud on the deep ocean floor during Silurian time. Then it crosses the great Katahdin granite pluton and ends in Silurian volcanic and sedimentary rocks near the dam.

Bouldery Till and the Katahdin Pluton

The Golden Road crosses the Katahdin pluton between the camps on Millinocket and Ambajejus Lakes and the area beyond the Telos Bridge—from mileposts 10 to 29. A variable thickness of sandy glacial till that contains boulders of Katahdin granite up to the size of houses covers the pluton. Bouldery moraines that rib the till-covered surface probably record stages in the melting of the last glacier. Bed-

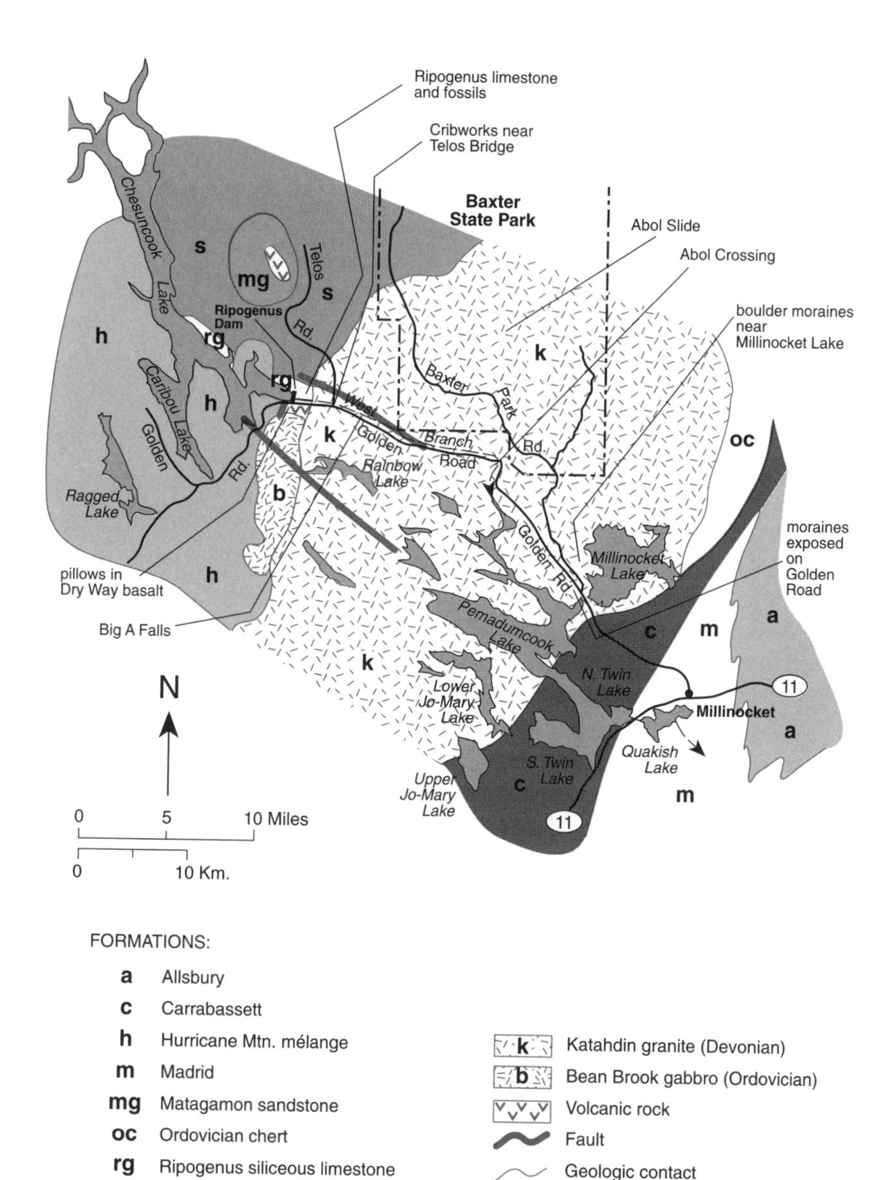

Geologic features along the Golden Road from Millinocket to Ripogenus Dam.

rock is exposed only along rivers and streams, where it commonly makes waterfalls and rapids.

Abol Crossing and the Great South Face of Katahdin

In September of 1846, Henry David Thoreau reached the mouth of Abol Stream in a wooden bateau poled up the West Branch. He was accompanied by loggers repairing booms and dams for the spring drive. His party camped near the present Abol Bridge Campground.

Thoreau intended to climb Katahdin, which he spelled "Ktaadn." He followed the trail along the Abol Slide to the Table Land, as had C. T. Jackson in 1835, during his first geological survey of Maine. Thoreau reached the Table Land at the top of the Abol Slide in dense clouds. The clouds, and fear that his companions might leave him behind, forced him down without reaching the summit.

The Abol Slide came down the mountain in 1816 and is still visible—a brown vertical gash near the left side of the mountain as seen from Abol Bridge. You can see other slide scars, especially on the smaller mountains west of Katahdin: the Owl, Barren Mountain, Mt. OJI, and Doubletop. Mt. OJI is named for slide scars that looked like those letters long ago, but continued sliding has nearly obliterated their shapes.

Boulder moraines and flat skyline of Mt. Katahdin from near Abol Bridge. Photo taken at milepost 17 in Township 2, Range 10. Boulders in foreground are Katahdin granite.

From the vantage point of Abol Bridge, the summit area of Katahdin appears almost level. This may have been the view that gave early geologists the idea that the top of Katahdin is a remnant of an old erosion surface that was thought to extend to Mt. Washington in New Hampshire. The top of Katahdin is not actually flat; it slopes down to the north. It is the exposed top of a granite pluton. The finely crystalline rock at the margin of the granite supports the summit of Katahdin.

The upper portion of the granite pluton, including all of the Summit facies and the upper several hundred feet of the Doubletop facies, is pink. The color is in the feldspar grains that contain minute grains of the iron oxide mineral hematite. The lower portion of the pluton is gray, because the feldspars lack the hematite inclusions.

The mountain summits west of Katahdin become progressively lower. This decrease in elevation reflects the original lens shape of the top of the pluton. The highest part of the pluton seen from here is South Peak, about 5,240 feet. The western limit of the Table Land near Thoreau Springs is about 4,600 feet; to the west the Owl is 3,730 feet, Barren 3,680 feet, and OJI 3,400 feet.

The elevations of the mountains west of Katahdin are not quite comparable to those on Katahdin, because the Summit facies, which is about 500 feet thick, has been eroded from the others. They consist entirely of the much less resistant Doubletop facies. That these western peaks are here at all must mean that the Summit facies has been eroded from them recently, at least in a geologic sense.

Big A and the Horserace Pluton

Big A is the pronounceable version of Big Ambejackmockamus Falls, the zigzag in the course of the West Branch near milepost 27. The Devonian West Branch fault controls the course of the river between Ripogenus Dam and Abol. The river jogs away from the road where it begins to follow fractures that are about at right angles to the West Branch fault.

In the early 1980s, the Great Northern Paper Company planned another dam on the West Branch at the Big A Falls for electrical generation. Some objected to the dam on geologic grounds, claiming that the West Branch fault would move when water impounded behind the dam seeped into it. Of course, that would break the dam and flood Millinocket. That was not a reasonable fear. The West Branch fault has not moved since Devonian time, some 380 million years ago, and seems most unlikely to move now, water or no water. Part of the Ripogenus Dam is on a branch of the same fault.

Outcrops along the Golden Road at Big A expose the Horserace diorite, which intruded along the West Branch fault. This rock contains more iron minerals than the Katahdin granite. They are black, and that explains the dark color of the diorite. The diorite magma may have formed as molten basalt mixed with molten granite.

Telos Bridge and the Cribworks

Numerous outcrops near the Telos Bridge near milepost 29 are some of the most accessible at the western margin of the Katahdin granite, which extends about a mile farther west. The rock is gray granite of the Doubletop facies, full of fractures and little faults. The granite is pink along some of the fractures. Evidently, hot water or steam circulated through those fractures, oxidizing the iron in the orthoclase feldspar to hematite, or red iron oxide.

Several faults trend northwest from within the Katahdin pluton into the enclosing older rocks. Some of them offset the western margin of the granite. The West Branch Penobscot River eroded its channel along some of those faults between Ripogenus Dam and Abol Bridge. The northeastern jogs in its course follow strong sets of parallel fractures that are at about right angles to the faults.

The section of the West Branch Penobscot River near Telos Bridge is now a popular stretch for white-water rafting and kayaking. During the old log drives, a dam was built across the small channel south of the bedrock island below the bridge to keep water and logs in the

Tilted slab of Katahdin granite at the Cribworks, West Branch.

main channel. Frames of logs filled with rocks—cribworks—supported part of the dam. It was removed in 1990 because it interfered with rafting.

Downstream from the bridge, on the north side of the river, is a platform of granite that is a popular place to watch the rafts go by. Look for the tilted slab of granite resting on an old cedar and an old pine. A crack already existed in the granite when the two trees took root many years ago. As they grew larger, their roots pried the slab up. As the trees swayed in the wind, they pried the slab up even more. It now stands almost on end.

Ripogenus Dam

Ripogenus Dam, or Rip Dam, is about 0.5 mile north of the Golden Road. It was built in 1915 to flush logs downstream to the paper mills in Millinocket. In the 1950s, about twenty years before the log drives ended, the dam was refitted to generate electricity for the same paper mills. The buildings near the dam at Pray's included homes for the workers on this project and their families, as well as a school and a store. The complex is now operated as housing for hunters, fishermen, archaeologists, and even occasional stray geologists.

The rocks below the dam are oceanic basalts, the Dry Way volcanic rocks. They contain pillow structures, which leave no doubt that the flows erupted underwater. They are oceanic crust that formed during Ordovician time.

The name Dry Way refers to one of two possible canoe routes used before the dam was built. The Wet Way went through the gorge on the north side of the dam; the Dry Way was a carry around the gorge, over the basalt outcrops.

The stream exploited the broken rock in the West Branch faults as it eroded its channel, accentuating them, especially below the middle of the dam. I should emphasize that these faults have not moved for a long time, probably not since Paleozoic time. They pose no conceivable threat to the dam.

The Ripogenus formation is exposed along the gorge near the north side of the dam, as well as along the Dry Way. In both areas, weathered basalt gives way to a thin, quartz-pebble conglomerate cemented with reddish iron oxide. Both the weathered basalt and the conglomerate suggest that the rocks were exposed to the atmosphere. Then they were again submerged, at first to a shallow depth when the limestone was deposited, then to greater depth, when the limy mud that became the upper Ripogenus formation was laid down. The Ripogenus limestone is in fact a sandstone with pods of fossiliferous limestone

Dry Way basalt on the right and Ripogenus formation on the left, exposed downstream of the dam. The concrete structure on the left is a sluice used to send logs from the lake into the river.

that make a discontinuous bedding. The orange cliffs above Ripogenus Lake are limestone with a thick crust of orange lichens.

Towering cliffs of the Fat Man's Woe volcanic rocks overlook the dirt road that leads from the north side of the dam. These are andesite lavas, the kind that typically erupt from volcanic chains that develop above a sinking slab of oceanic crust. These probably erupted during the beginning of the Acadian mountain-building event, while the floor of the Iapetus Ocean was sinking beneath what is now Maine.

Outcrops across from the power station, about a mile east of Ripogenus Dam, expose the base of the Katahdin granite pluton where it intruded the Fat Man's Woe volcanic rocks. The contact slopes down toward the east, toward the center of the pluton, suggesting that the base of the intrusion is as lenticular as the top. If you project this contact to the west, it passes through cliffs high on Fat Man's Woe, which suggests that before erosion the Katahdin granite probably extended beyond its present western margin.

Near the power station is the remnant of a large pile of crushed rock removed from the tunnel that carries water from Ripogenus Dam to the electrical generators. The rock is a curious interbedded black-

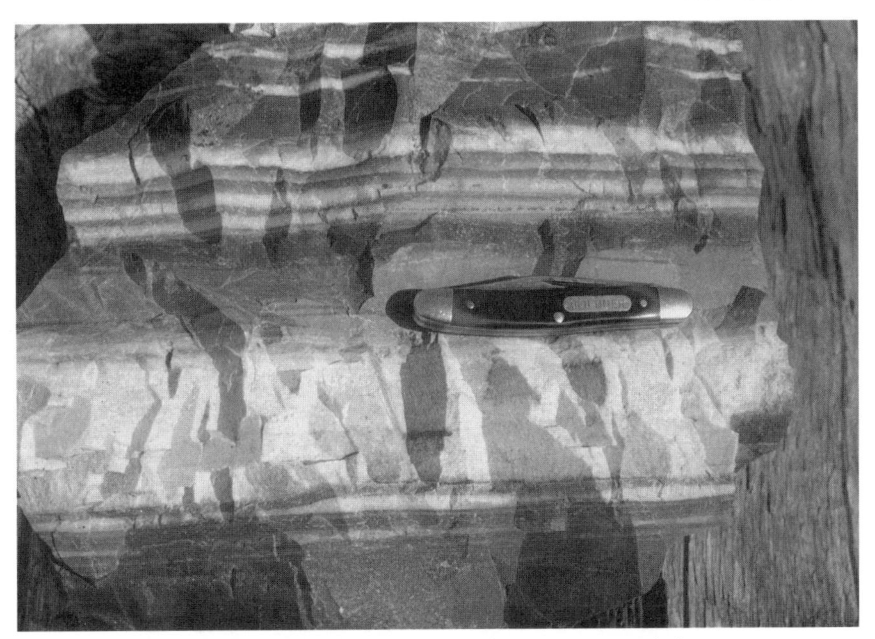

Banded argillite of the Ripogenus formation. Local chimney and fireplace builders call this rock the "Rip Dam Striper."

and-white argillite. It may represent the upper, thin-bedded calcareous unit in the Ripogenus formation that was baked by the Katahdin pluton. The heat probably drove off the carbon dioxide in the rock, compacting it somewhat. I have never found an outcrop of this rock and apparently the glaciers did not either, because it does not occur in till or outwash deposits. The banded argillite is found over much of Maine in rock gardens and so forth, taken as souvenirs from the pile near the Rip Dam power plant.

West Branch Raft Trip

Henry David Thoreau may have been one of the first tourists to do the West Branch trip. In 1846 he came up the West Branch in a 300-pound bateau and was given a ride up one of the falls. He camped at Abol Bridge during his attempt to climb Katahdin.

Rafting companies generally run trips on the Kennebec and West Branch Penobscot Rivers. The Kennebec River is in a deep gorge, with little to see but its rock walls, while the West Branch River offers many varied views of Katahdin and other mountains. The Kennebec River has rushing, turbulent waters along nearly the whole trip, while the West Branch River has invigorating stretches between long reaches of flat water.

If you start near the power station, the whole West Branch raft trip is in the Katahdin granite, but the rock is not all the same. Along the first 10 miles or so, from the power station to Abol Bridge, the West Branch follows a fault; it explains some of the long straight stretches of the river. At the Cribworks, beyond the Telos Bridge, the river turns sharply to the left, following a strong set of fractures oriented at right angles to the West Branch fault. Another major jog at Big A Falls also follows fractures perpendicular to the fault. At about that point watch for much darker rock, the Horserace diorite. It is a small pluton that intruded along the West Branch fault.

A large intrusion into the Katahdin granite, the Debsconeag granodiorite, begins about at the Abol Bridge and continues downstream to Debsconeag Falls. This rock is darker than the Katahdin granite, but lighter than the Horserace diorite upstream. Both of the smaller plutons are Devonian in age and likely were involved in the formation of the large Katahdin pluton.

Roaring Brook Road
Togue Pond Gate—Roaring Brook Campground

11 miles

When Percival Baxter donated the lands of the park to the people of Maine, he expressly forbade any modernization of the existing roads. They are now just as they were in the 1930s, narrow dirt tracks for which the posted speed limit of 20 miles an hour is too fast.

Togue Pond Esker

The road from Millinocket that links with the Golden Road at Millinocket Lake leads to the south entrance to Baxter State Park. It is locally known as the State Road. At the end of the pavement are two pretty lakes called Upper and Lower Togue Ponds; they are large kettle ponds in glacial outwash. A togue is a lake trout. The Girl Scout camp on Lower Togue Pond is the site of a Civilian Conservation Corps camp of the 1930s.

Baxter Park Road is on a segment of a large esker along several miles from the south gate toward Abol Campground. The sand and gravel are well drained, making a dry roadbed even during rainstorms.

Below the 20- to 50-foot-high esker are numerous swamps and lakes. The high road on an esker is much nicer than the low road through the swamps and between the lakes.

The esker extends eastward around Millinocket Lake and then follows Millinocket Stream. Like all eskers in Maine, this one stays in stream valleys. It crosses from one drainage into another through the lowest possible gap. South of Millinocket the Katahdin-to-Cherryfield esker runs up the valley of Nollesemic Stream, around Nollesemic

Features formed by glaciers near the southern end of Baxter State Park.

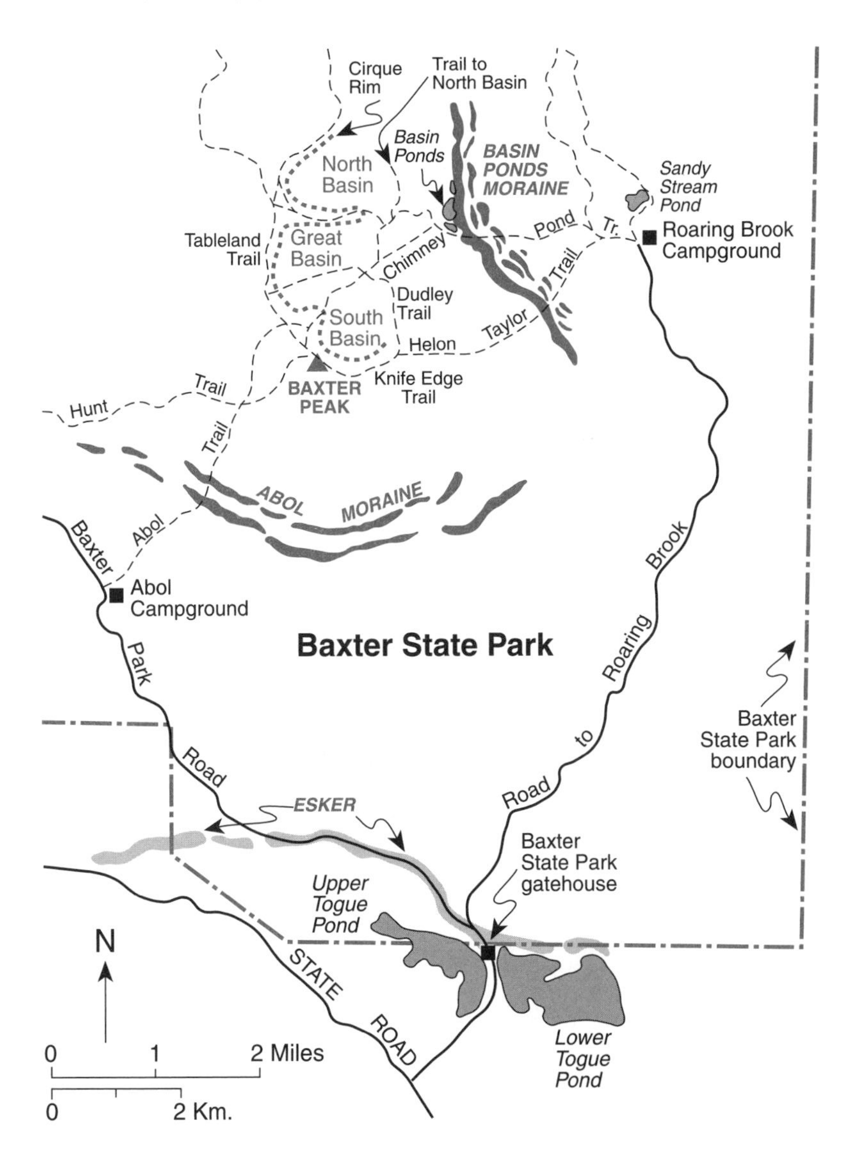

Pond, and follows swampy ground into the head of Sam Ayers Brook, which it follows to the Penobscot River. It crosses Interstate 95 at milepost 207. It continues southeastward, crosses the Penobscot River near Lincoln, and ends in a glacial marine delta near Cherryfield.

Maine eskers must have formed in streams that flowed in tunnels under the ice. They never drape haphazardly over hills as they might had they flowed over the surface of the ice, as some have suggested. And the layers of sand and gravel in them are broken only near their edges. If they were deposited in surface streams on the glacier and let down to the ground when the ice melted, the beds of sand and gravel would be broken throughout.

Avalanche Brook

The first major clearing along Roaring Brook Road is at Avalanche Brook and its campground. Avalanche Brook and Keep Ridge above it were used for some of the early climbs of Katahdin. The Reverend Marcus Keep of Bangor was the guide for many of these early climbs and had cleared a trail known as Keep Path in 1846.

The road to Roaring Brook Campground climbs about 1,000 feet, with several steep pitches. Most of these are over bedrock thinly covered with glacial till and boulders. Small pockets of bouldery meltwater deposits exist in a few places. Just before it descends into the campground, the road passes through a rather deep valley, apparently eroded by glacial meltwater from ice located in the valley of Roaring Brook.

Helon Taylor Trail, the Knife Edge, and Baxter Peak

10 miles round-trip

This trip is a strenuous hike of about 10 miles, depending on the exact route, with a climb of more than 4,000 feet. The best route to observe changes in the granite is from Roaring Brook Campground to Pamola Peak by the Helon Taylor Trail. The hike up the Helon Taylor Trail, across the Knife Edge to the summit at Baxter Peak, and down to Roaring Brook Campground again requires about ten hours, especially if you stop to look at the rocks.

About a mile and a half from Roaring Brook, the trail crosses bouldery ridges of the Basin Ponds moraine. Above the moraine the trail crosses outcrops of the coarsely crystalline Doubletop facies of the Katahdin granite pluton. Many outcrops expose rotten granite in which the grains of feldspar and quartz have begun to separate, forming loose debris of feldspar and quartz called grus. The Doubletop granite becomes finer grained upward, and finally grades into granite of the Chimney facies in which intergrowths of quartz and feldspar, visible only with a magnifier, make up as much as 60 percent of the rock.

The large dark spots in the upper part of the Chimney facies are cavities filled with minerals such as black tourmaline and yellowish green epidote. They were bubbles of steam in the magma that were trapped and crystallized into rock. Water and steam circulating through the cooling rock deposited the minerals in the cavities. The appearance of open cavities, only partly filled with crystals, defines the base of the Summit facies, which extends from about 4,500 feet to the summits of Pamola Peak at 4,902 feet and Baxter Peak at 5,268 feet. The facies is at least 700 feet thick.

Open cavities in granite, or cavities filled with minerals, are good evidence that the magma crystallized at shallow depth, probably within a few thousand feet of the surface. Their presence in the upper part of the Katahdin granite adds weight to the argument that it was the

Partly filled cavities of the Summit facies of the Katahdin granite.

magma chamber for the Traveler rhyolite. Volcanic rocks almost certainly erupt from shallow magma chambers.

The Helon Taylor Trail ends at the summit of Pamola Peak, named for the mythical Indian god of Katahdin—part man, moose, and eagle. The Dudley Trail leads down to Chimney Pond, and the Knife Edge extends another mile to Baxter Peak. The Knife Edge is an arete, a narrow ridge between two glaciated valleys. Glaciers in the South Basin cirque eroded the north side while continental ice eroded the great south wall of Katahdin south of the arete. The crest is some 3 to 8 feet wide—a knife edge, indeed.

Baxter Peak, named Monument Peak before this region became a park, is little more than a bulge on the flat upper surface of the Katahdin pluton. The word *monument* referred to a survey line laid across Maine in 1825. The monument line is the control for the survey of the townships that cover a large part of northern Maine. This survey was begun on the eastern border of Maine, at the boundary with New Brunswick. The line was run due west, taking into account the magnetic declination. The townships surveyed north and south of the monument line are the only towns in Maine with true north-to-south and east-to-west boundaries. All other town lines are based on magnetic north, not on the direction to the pole star.

Joseph Norris and his surveying party carried the line across the Table Land of Katahdin, about 2 miles north of Baxter Peak, during a November snowstorm, then stopped work for the year. Out of supplies, they spent a miserable night on the Table Land. The next day they descended the Abol Slide to the West Branch, where tents, canoes, and food were waiting for them. They just made it to Bangor before the Penobscot River froze for the winter. The next year they started on Chesuncook Lake, intending to hit the line from the west, but they missed it by about a half-mile. The jog that connects the lines is near Doubletop Mountain. All the townships east of Doubletop Mountain are about a half-mile longer, north to south, than those to the west.

A pile of boulders at the very top of Baxter Peak raises its elevation to exactly a mile. For about 100 years the elevation of Baxter Peak was given as 5,267 feet; then in 1980 it was given as 5,268 feet. The crust of northern Maine is still rebounding from the weight of the last ice sheet, so that might explain the difference. The elevations of mountains are rounded up or down on a map to the nearest whole number. So a small change in actual elevation might change the map elevation as much as a foot.

To the east of the summit is the nearly vertical 2,000-foot drop into the South Basin cirque and to Chimney Pond, which is a tarn—a lake basin in bedrock scooped out by a valley glacier. Among the loose boulders of granite at the summit are a few erratics of sandstone and volcanic rocks carried in the continental ice that covered Mt. Katahdin.

Look west from Baxter Peak to see the numerous large slide scars on Doubletop Mountain and Mt. Coe. Beyond the edge of the Table Land is a large, swampy area called the Klondike. The curious bands of dead spruce trees are visible on Mt. Coe and other mountains west of the Klondike. Between Katahdin and Hamlin Peak to the north is a broad low area called the Saddle. The low trees on either side of the trail are a kind of timberline growth called krummholtz. They grow along the ground, very slowly. Some are one hundred years old, less than 3 feet high, with branches 20 feet long. The howling winds that blow across the Table Land prevent the trees from growing higher.

The Saddle exists partly because the Summit facies that protects much of the high country is gone, exposing the more coarsely crystalline and much less resistant Doubletop facies of the Katahdin granite. The granite along the trail down to Chimney Pond is faulted and deeply weathered and rotten. Glaciers passing over the mountain probably had an easy time with this rotten rock and eroded the Saddle to near its present elevation. Continued erosion in the Saddle will eventually separate Baxter and Hamlin Peaks.

Mountain summits to the northeast far beyond Hamlin Peak belong to the Traveler Range. Bedrock there is beds of volcanic ash that erupted from the same magma chamber that became the Katahdin granite.

Chimney Pond Trail

6 miles

The Chimney Pond Trail crosses a major belt of moraines deposited from valley glaciers that flowed out of the large cirques on the east side of Katahdin. The total length of this hike is about 6 miles, with a climb of about 2,000 feet. The short hike to Sandy Stream Pond from the Roaring Brook Campground is a good substitute. You can easily see the major moraines and cirques from the pond, and usually a few moose feeding on the aquatic plants. The moose put their heads underwater and come up, snorting and blowing, to chew.

About a mile and a half along the Chimney Pond Trail, a short spur to the north leads to a small moraine with no tree cover. Most of the rock fragments are Katahdin granite; about 20 or 30 percent are erratics carried in from a distance. The continental glacier left this moraine. To the north you can see a broadly gouged valley nicely framing the Traveler Range of volcanic rocks. A stream originally eroded that valley, then a glacier used its embedded rock fragments to gouge it out.

At 1.75 miles from Roaring Brook Campground, the trail crosses the Basin Ponds moraine, one of the largest in New England. It is several miles long, and large enough to dam several ponds on its uphill side. The rock fragments in it are mostly Katahdin granite, which leads geologists to believe that the ice came from the cirques to the west. That the moraine continues far south of the cirques may mean that the valley glaciers flowed to the Basin Ponds and merged with or perhaps flowed onto a remnant of the continental ice sheet, which dragged the moraine south.

At 2.3 miles from Roaring Brook Campground, a trail to the right leads to the mouth of the North Basin cirque. Another trail leads from North Basin to the South Basin and Chimney Pond. The floor of North Basin is above timberline, so several small moraine ridges are quite obvious.

The pile of boulders at the very head of the North Basin cirque, below the headwall, is a rock glacier. It might be moving. Years ago, geologists considered rock glaciers the last remnants of melted valley glaciers. Unfortunately for that idea, many rock glaciers exist in valleys that never held ice.

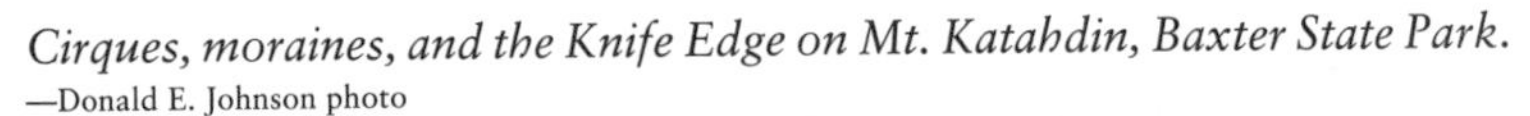

Cirques, moraines, and the Knife Edge on Mt. Katahdin, Baxter State Park.
—Donald E. Johnson photo

It seems more likely that a pile of boulders fills with ice for the same reason that an ice cave does: Air can sink into the crevices between boulders only if it is colder, therefore denser, than the air already there. That happens only on the coldest days of winter. So the crevices between boulders fill with ice, and the insulating properties of the rocks keep it through the summer. The average temperature within the pile of boulders is much lower than the mean annual temperature outside. If the pile of boulders is at least 50 feet high, it exerts enough pressure on the ice at the base to make it flow exactly as it would in an ordinary glacier. So rock glaciers do move, although much more slowly than ordinary glaciers.

The North Basin and the great South Basin cirques are about a mile apart. Above the cirque floor at Chimney Pond towers the 2,000-foot headwall, terminated on the left by the Knife Edge and in the middle by Baxter Peak. The heap of boulders that brings the summit to an even mile is clearly visible.

The smooth bedrock surfaces in the lower part of the cirque are glacially polished, presumably because the last ice in this basin was a valley glacier. The straight gullies that streak the slopes above this smoothed portion of the headwall are avalanche tracks.

Maine 159 and Grand Lake Road
Patten—Baxter State Park

30 miles

Lumberman's Museum

The Lumberman's Museum in Patten contains reconstructed lumber camps and displays of tools and equipment. They include the redoubtable Lombard log hauler, which was a railroad steam engine fitted with skis in front and caterpillar treads in the rear.

Grand Lake Road

Grand Lake Road is paved nearly to the park. Between Shin Pond and the East Branch Penobscot, it follows a Civilian Conservation Corps road built in the 1930s. The road between Shin Pond and the East Branch is known as the Grand Lake Road for Grand Lake Matagamon.

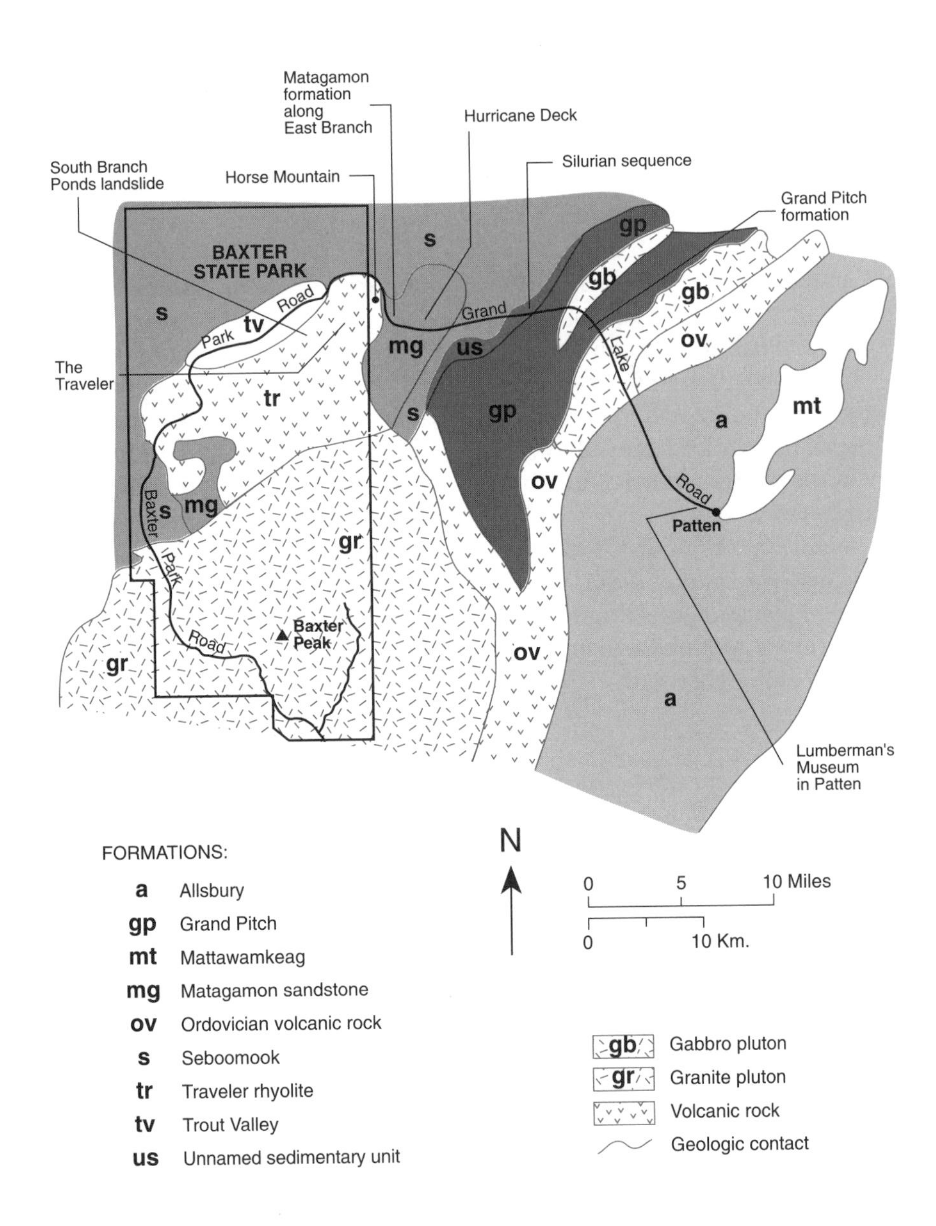

Geologic features of Baxter State Park and its northern approaches.

Rocks along the Grand Lake Road are as variable and interesting as on any 30-mile stretch in Maine. Slates of the Allsbury formation of Silurian age, part of the Central Maine slate belt, extend to Shin Pond. Rising east of Shin Pond are the conical peaks of the Mt. Chase volcanic belt. These are Ordovician in age and erupted in a volcanic chain associated with the closing of the Iapetus Ocean. The peaks owe their shapes entirely to erosion, not to the volcanic origin of the rocks in them.

West of Shin Pond, the Grand Pitch formation is exposed along several streambeds. It is Cambrian in age, well dated by fossils. The Grand Pitch formation is named for rapids on the Seboeis River, reached by a path that follows the west bank of the river south of the Grand Lake Road. Volcanic ash supplied part of the sediment for the turbidites of the Grand Pitch formation.

Ordovician gabbros and other igneous rocks exposed near the Grand Lake Road may have crystallized from the same magma that supplied the volcanic rock at Chase Mountain, nearby. They are exposed on Sugarloaf Mountain and Green Mountain.

At Hay Lake, the Grand Lake Road crosses into the Devonian sedimentary rocks that are the main focus of this field trip. South of Hay Lake, along the road to the Bowlin Pond, are fossiliferous Silurian sedimentary rocks.

Matagamon Formation

For many miles the Grand Lake Road is absolutely straight, laid out from east to west. West of Hay Lake it begins to wind noticeably through large outcrops of the Matagamon formation. This dark sandstone was deposited as a delta building into the deep ocean from the Acadian Mountains that rose during the collision of Avalonia. It is sparsely fossiliferous, with brachiopods being the most common. They show that it was deposited during Devonian time, perhaps at the same time as the Tarratine formation to the southwest.

Watch near the first hairpin turn where the Traveler Range is visible for a turnout south of the road, the Hurricane Deck. From this vantage point it is possible to see features that are lost close-up. The mountains to the west all have long north slopes of about 20 degrees. These are exposed beds of ash that erupted explosively from the Traveler volcano. The much steeper south slopes on these mountains are eroded along columns that broke the ash vertically.

The highest mountain in view is Traveler, 3,500 feet. It is the highest mountain in Maine made of volcanic rock, perhaps the highest in New England, possibly in the eastern United States. The Traveler and

North and South Traveler Mountains rise above tree line and support arctic plants similar to those on Katahdin and in the arctic.

Garlands of boulders near the summit of Traveler are solifluction lobes, or, if you prefer, soil flowage lobes. They form during spring and summer in regions of permanently frozen ground when the surface soil thaws while that at depth remains frozen. The sloppy surface soil slides on its icy base, making curved bands of soil and boulders.

East Branch

In this part of Maine, where people like to use as few words as possible in conversation, East Branch means East Branch Penobscot. The words "of the" do not appear, nor does "River." More often it would be called just the East Branch. The Penobscot begins where the East and West Branches merge near Medway. In the last of his three trips to the Maine woods, in 1857, Henry David Thoreau finished the canoe portion of the journey with a trip down the East Branch, passing by the site of the present bridge in late July.

Horse Mountain

Horse Mountain, near the north entrance of the park, was apparently named for the hind end of a horse. Charles Hitchcock, in conducting a geological survey of Maine in 1861, referred to it as the beautiful mountain with the inelegant name. Along the road south of Horse Mountain are several outcrops of Matagamon sandstone that dip toward the west, underneath cliffs of Traveler rhyolite on Horse Mountain; the sandstone is older than the rhyolite.

At the base of the high cliff are large boulders of Traveler rhyolite, fragments of large columns that fell from the cliffs above, where you can see the bottoms of intact columns. Geologists divide the Traveler rhyolite into the lower Pogy member, exposed here, and the upper Black Cat member. The Pogy member probably erupted from the upper regions of the magma chamber, where the magma was cool enough for feldspar and quartz to crystallize. These crystals were erupted with the ash that makes the Pogy member. Later eruptions came from the deeper and hotter portions of the same magma chamber and produced the Black Cat member, which contains a few crystals of feldspar, none of quartz. The two members of the Traveler rhyolite represent an inverted version of the original Katahdin magma chamber.

The Traveler rhyolite is essentially identical with the rhyolite of the Kineo flint to the southwest. Both rocks are dark gray, with small crystals of quartz and other minerals. Both rocks weather to a chalky white color. This weathered surface is less than a millimeter thick. I know of no record of the Traveler volcanic rock being used for stone implements by early Indians as was the Kineo flint.

The contact between the Matagamon formation and the Traveler rhyolite is exposed at the base of the cliff at the top of the talus pile of broken columns. The brave or immune geologists who have waded through the luxuriant growth of poison ivy to see it report that the lower beds of ash were deposited in water and include pieces of the underlying sandstone. Another bed or two of sandstone was deposited, and then sea level dropped and the rest of the Traveler ash beds were deposited above sea level.

South Branch Ponds

Ice scoured the long basin that holds Upper and Lower South Branch Ponds during the last ice age. This valley continues south through the park, past Russell Pond and Roaring Brook Campgrounds. The length and general straightness of the valley lead me to believe that rivers and then glaciers eroded it along a fault, although no fault appears on the geologic map.

At some time after the ice melted, a large landslide or debris flow moved from the valley of Howe Brook to the east, into the lake, dumping a deposit that divided the single lake into two. It is not known if this was a single flow or several, but a flow did move fast enough to climb a short way up the mountain on the far side of the pond. This is the largest landslide deposit known in Maine.

Trout Valley Formation

South Branch Ponds Brook eroded a gorge that drains north to Trout Brook. The Trout Valley formation was originally deposited on the floodplains of rivers draining the recently erupted volcanic mountains and represents the first nonmarine sedimentary rock to form in this area. Unlike most of the sedimentary rocks in Maine, these are not even slightly metamorphosed. Rocks in the Trout Valley formation are little more than particles of sediment glued together.

The walk down this brook takes about three hours, and includes wading. About a mile north of the South Branch Ponds Campground, the South Branch Falls trail leads to the west. Outcrops of the Traveler rhyolite are well exposed at South Branch Falls. Prominent layers of collapsed lumps of pumice are about at right angles to well-developed columns and represent the only evidence of bedding. The boulders in the stream channel are quite slippery, but I always find one of those walking sticks the beavers make and get down the brook without falling.

About 100 yards downstream, the Traveler rhyolite gives way to a conglomerate at the base of the Trout Valley formation. It was deposited from a debris flow, or perhaps from volcanic mudflows. The conglomerate consists of rounded and broken columns of Traveler rhyolite, cemented with ash. A few hundred yards farther downstream, the Trout Valley formation is weathered to a reddish orange. Because

Geologic features along the walk and wade down South Branch Ponds Brook.

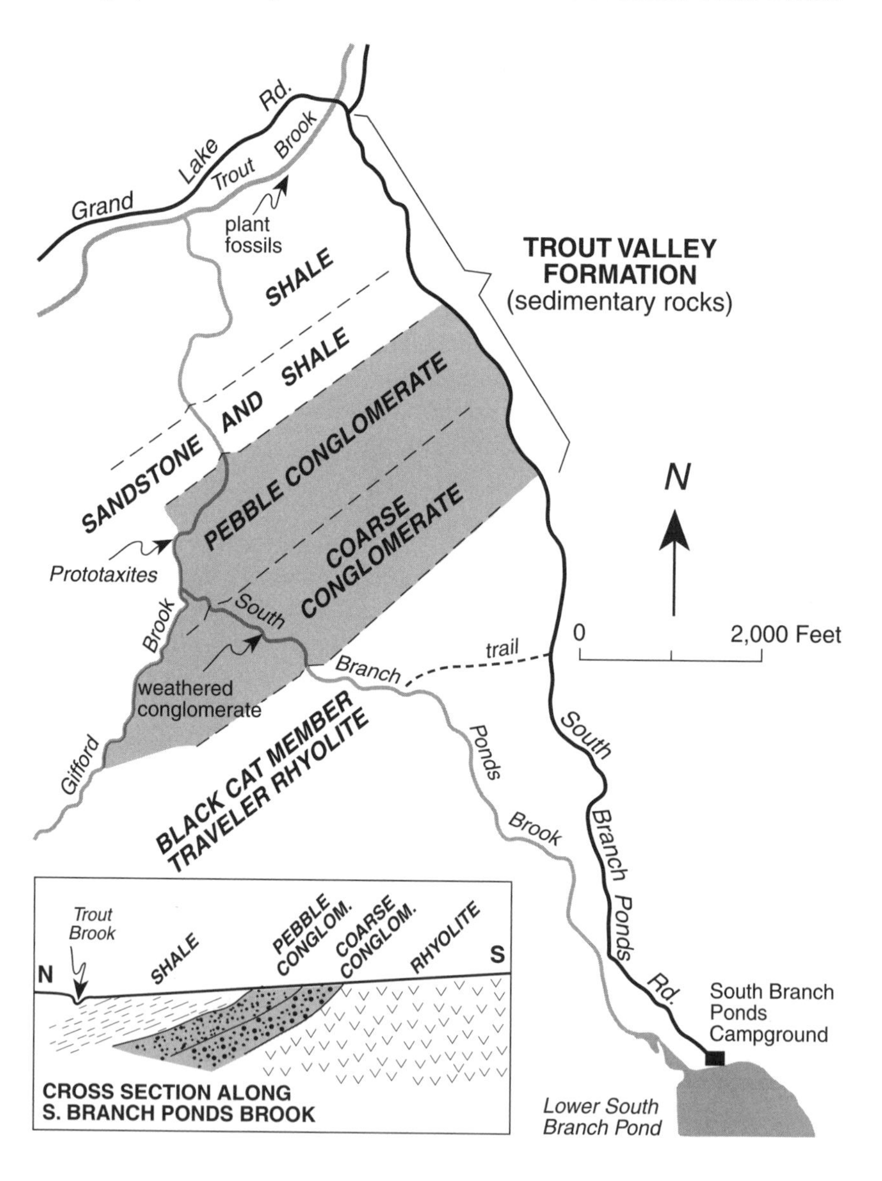

Conglomerate of the Trout Valley formation.

the overlying rocks are not weathered, geologists believe the weathering happened during Devonian time, before the rocks above were laid down.

Small faults offset many of the boulders in the weathered conglomerate a few inches. The Acadian mountain-building event seems an unlikely culprit because it ended before the conglomerate was deposited.

Outcrops near the junction with Gifford Brook expose conglomerate made of smaller rock fragments, with layers of sandstone. The bedding dips gently downstream. As you follow these rocks downstream, you are climbing up through the formation into progressively younger rocks higher in the Trout Valley formation.

A high bank at a right-angle bend in the channel marks the middle portion of the Trout Valley formation, which consists of sandstone and pebbly sandstone. Iron oxide and iron carbonate cement some beds, which make little ledges. Near the water line is a fossil plant named *Prototaxites,* a curious nodule of black chert that may have formed from the stem of a giant Devonian mushroom. Some specimens of *Prototaxites* are 3 feet in diameter and 5 feet long. Some mushroom.

Pebble conglomerate with sandstone beds in Trout Valley formation. The dark lump over the man's head is a Prototaxites, *possibly from the stem of a Devonian mushroom.*

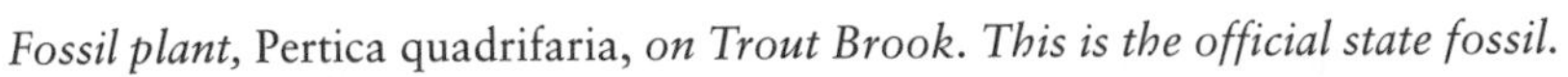

Fossil plant, Pertica quadrifaria, *on Trout Brook. This is the official state fossil.*

Shale exposed along Trout Brook is in the upper part of the Trout Valley formation. It outcrops occasionally in the last mile of South Branch Ponds Brook and downstream along Trout Brook to the bridge known as the Crossing. Black dikes cut across the shale in several places. It is surprising to find these dikes because all tectonic activity is supposed to have been over by the time the Trout Valley formation was deposited.

Abundant fossils of Devonian land plants in the Trout Valley shale consist of reddish brown stems and fruiting bodies. The living plants were no more than a few feet high and lived in swampy ground, probably on river floodplains. One of these plants, *Pertica quadrifaria*, found a couple of miles upstream on Trout Brook, is the state fossil.

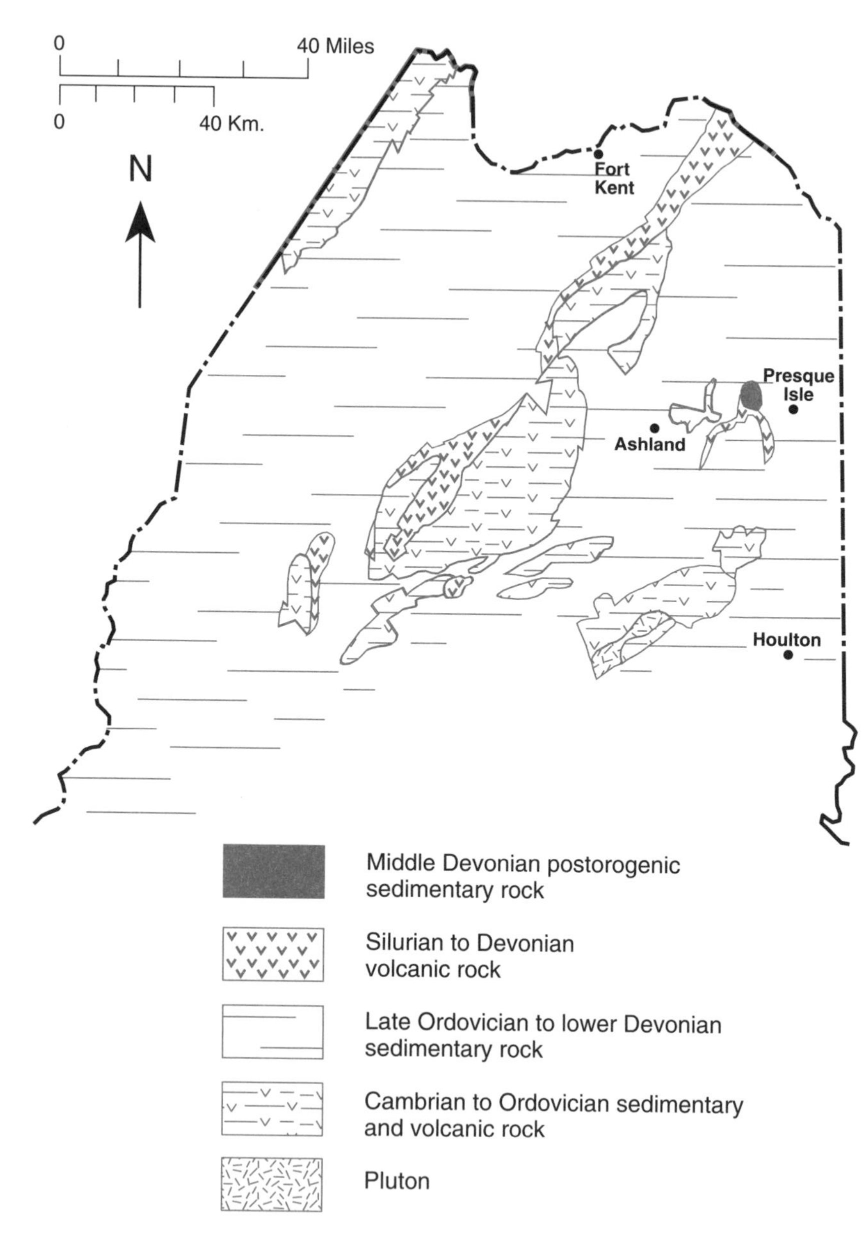

Geologic features of northern Maine.

Northern Maine and The County

Northern Maine includes all the land drained by the St. John River and its tributaries, such as the Allagash, the Fish, the Aroostook, and the Meduxnekeag Rivers. The area includes nearly all of Aroostook County and the northernmost parts of Penobscot, Piscataquis, and Somerset Counties. Aroostook County is known by most Mainers as simply "The County," also by a more recent Chamber of Commerce designation as "the Crown of Maine."

The Aroostook War

The Aroostook War, between 1839 and 1842, was our last with Britain, and perhaps the only bloodless war in history. Troops from both sides marched around Aroostook County in 1839, with General Winfield Scott credited with keeping calm in both armies. The position of the border between Maine and Canada had been described in the treaty between Britain and the United States at the end of the Revolution. However, the lack of adequate maps or of people with any knowledge of interior Maine put the border in almost continuous dispute between 1783 and 1842. The Treaty of Paris of 1783 described the border as extending due north from the most distant headwater of the St. Croix River to the watershed between rivers flowing down the St. Lawrence and those flowing into the Atlantic. The major difference between the present border and that described in the 1783 treaty is in the northern portion of this territory, in westernmost New Brunswick and adjacent Québec.

U.S. 2A, between Houlton and Macwahoc, is a paved version of a road built to get American troops to Aroostook County in a hurry and is still known as the "Military Road." Another road was begun in 1838 to speed troops to Calais from Bangor, but it was not finished until 1857 as the route of a mail and passenger stagecoach called the Air Line, a name still applied to the road, now Maine 9.

The present northern and eastern border of Maine was set by the Webster-Ashburton Treaty in 1842. Lord Ashburton and Daniel Webster oversaw the setting of the boundary monuments, about one every mile, along the new border.

Fort Kent, at the mouth of the Fish River, was built during the Aroostook War.

Boundary marker number 1 in Amity and Richmond, New Brunswick.

French Settlers in Aroostook County

The French explored the coast of Maine as early as 1604, but the first French settlers were mostly Catholic priests ministering to the Indians. Between 1675 and 1763 there were five French and Indian Wars in which Maine and Mainers were heavily involved. Most Indians in Maine sided with the French in these conflicts, since they had been generally poorly treated by the British and the American colonists. A great French fort at Louisburg in Nova Scotia was captured in 1745 by a force of colonists made up mostly of Mainers. The remaining French settlers living in what they called Acadia, after refusing to swear allegiance to the British king, were dispersed over a number of the English colonies.

Some Acadians settled near Fredricton, New Brunswick, in about 1759, but they were scattered again after the Revolution when Loyalists arrived in the region in force. A group of these Acadians were resettled in northern Maine in 1785, along the St. John River, the first settlement being at Grand Isle. Other Acadians were later resettled in Louisiana, where they became Cajuns.

New Crops in The County

Except for an unfortunate experience with sugar beets in the 1960s, commercial crops until recently have been almost entirely potatoes. Two grain crops, oats and barley, now alternate with potatoes, even in the same field. In the late summer, the golden fields of grain are set off by the darker greens of the potato stands. In addition, grain crops are rotated annually in the same field with potatoes. The third large crop in Aroostook County is broccoli. Broccoli can be grown on almost any kind of soil and the bluish green stands of this crop are likely to be found almost anywhere. Much of the broccoli sold in New England now comes from The County and probably accounts for the popularity of cream of broccoli soup in many restaurants.

A Shortage of Granite

Northern Maine is slate country. The few small bodies of granite and other intrusive igneous rocks, such as those at Deboullie Mountain, Priestly Mountain, Seboeis Lake, and Rockebema Lake, hardly compare with the many plutons in the coastal, central, and mountainous regions of the state. Intrusive igneous rocks underlie about 10 percent of the other parts of Maine but less than 1 percent of the northern part.

Slates of the Seboomook Formation

The most widespread rocks in northern Maine are slate and sandstone that belong to the Seboomook formation. All were deposited in seawater during early Devonian time, between about 400 and 380 million years ago. Indeed, this region is known as the Northern Maine slate belt, or the Devonian slate belt.

The rocks of the slate belt are part of the Connecticut Valley–Gaspé synclinorium. A synclinorium is a large and complex structure, a giant syncline with many little anticlines and synclines superimposed on it. Some evidence suggests that the Connecticut Valley–Gaspé structure follows a rift that opened after the Taconic mountain-building event. The Connecticut River follows the edge of it along the border between Vermont and New Hampshire. The principal rock there is called the Littleton formation. It is comparable in age and rock type to the Seboomook formation.

Cambrian, Ordovician, and Silurian Rocks

Rocks older than the Devonian Seboomook formation occur in three areas of northern Maine, two of them along major crustal arches. In the northwesternmost part of Maine is an extension of the Green Mountain–Sutton Mountain anticline, an elaborate arch that raises rocks of Cambrian and Ordovician age to the surface. Rocks in this belt are mainly remnants of old oceanic crust that collided with North America during the Taconic mountain-building event of Ordovician time. A characteristic rock associated with the Taconic event is a tectonic mélange, a mixture of mud and other rocks scraped off the sinking oceanic crust into an oceanic trench and stirred into a chaotic mess. The St. Daniel formation is such a rock; it appears along the Québec border. The Depot Mountain formation is another widespread older rock unit in northwestern Maine. It consists mostly of mudstones deposited during Ordovician and Silurian time, with small amounts of volcanic rocks.

The Munsungun-Winterville anticline is southeast of the Green Mountain anticline. It raises volcanic rocks of the Winterville formation to the surface. They erupted during Ordovician time, as part of the Taconic mountain-building event. The volcanoes were almost certainly a chain parallel to an oceanic trench, similar to the modern Aleutian volcanoes that parallel the Aleutian trench. The Munsungun-Winterville structure is a continuation of the Boundary Mountain anticlinorium to the southwest that in turn is an extension of the

Bronson Hill terrane of New Hampshire and Massachusetts. Other older rocks in this belt include the Madawaska Lake formation, mudstones deposited during Ordovician time.

Eastern Aroostook County has some limestone that was deposited during Ordovician and Silurian time, the Carys Mills formation, and during Silurian time, the New Sweden formation. These rocks are exposed in the Aroostook-Matapedia anticline.

Low-Grade Metamorphism

Mudstone is just a step away from slate. Except for rather weakly developed zones of baked hornfels around the few igneous intrusions, the sedimentary rocks of northern Maine have escaped serious metamorphism. Evidently, they have never been deeply buried.

Fossils survive in the sedimentary rocks of northern Maine because they generally escaped metamorphism. Some sedimentary rocks are very fossiliferous—most notably the Mapleton and Frenchville formations, which contain some rare fossils of middle Devonian plants. The sedimentary rocks of northern Maine are the only ones in New England that might conceivably contain crude oil.

Northern Maine Ice Cap

Near the end of the last ice age, about 13,000 years ago, an ice cap covered northern Maine. Earlier, an ice sheet that started in central Canada covered all of Maine with as much as 5,000 feet of ice; it reached nearly to the edge of the continental shelf at Georges Bank.

As the ice melted, its margin in southern and central Maine floated in the rising sea, which reached a short distance up the East Branch Penobscot beyond Medway, within about 20 miles of The County. Meanwhile, the ice near the mouth of the St. Lawrence River also floated. Seawater flooding the St. Lawrence Valley isolated the ice in Maine and southern Québec from the rest of the great ice sheet, forming a separate ice cap.

From about 20,000 years ago until about 13,000 years ago, the ice in Maine flowed south. Then, after the Maine ice cap was isolated, the ice near the Canadian border began to flow north, to the St. Lawrence River. Many outcrops in northern Maine show glacial striations that record ice flowing south, then north. Meanwhile, the ice in southern Aroostook County flowed southward toward the coast.

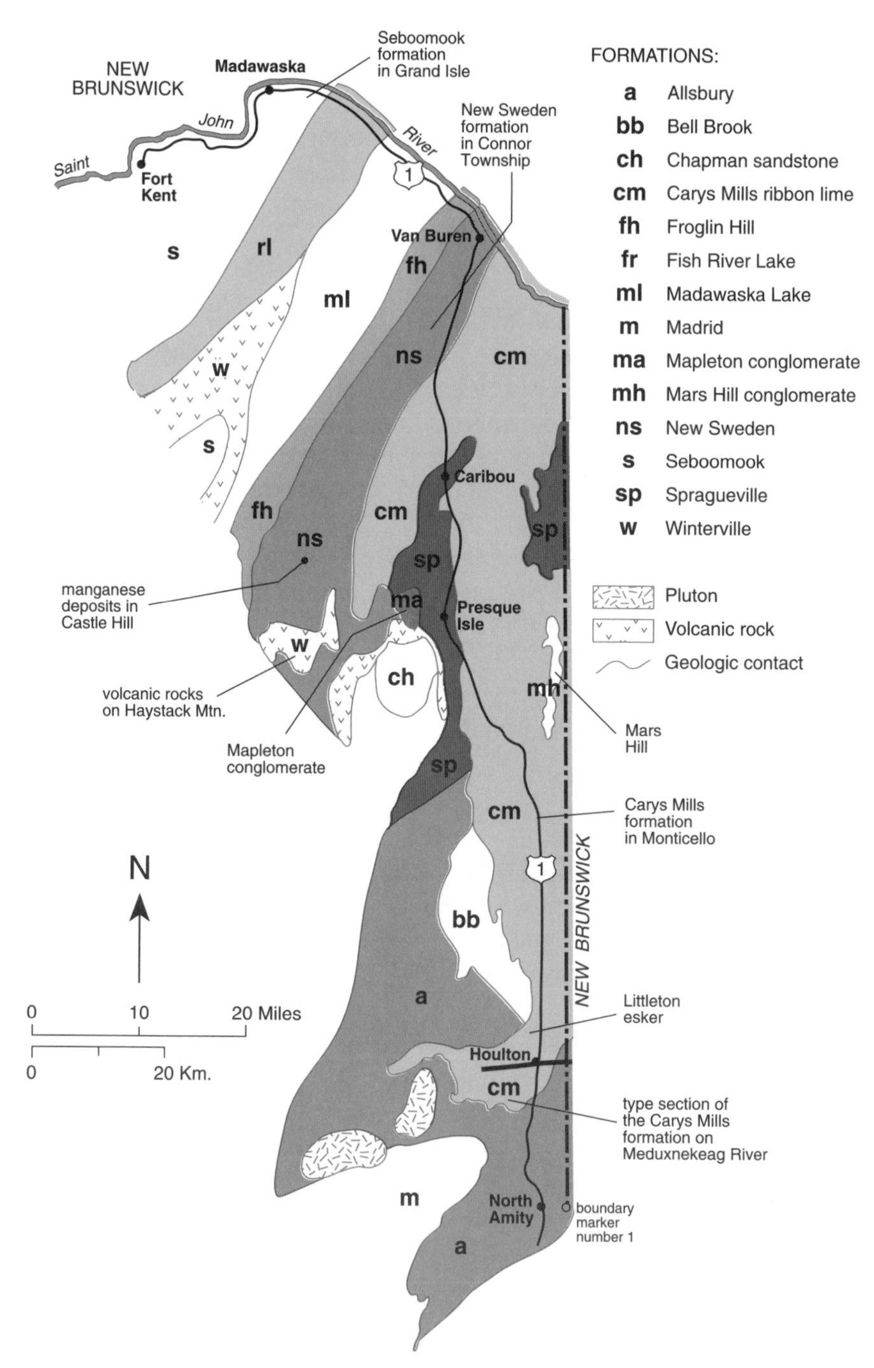

Geologic features along U.S. 1 between Houlton and Fort Kent.

U.S. 1
Houlton—Fort Kent
122 miles

U.S. 1 was strung together from a number of state and local roads in 1925. Its northern end is Fort Kent; its southern end is Key West, a distance of more than 2,500 miles. Nearly a fifth, 470 miles, of U.S. 1 is in Maine.

Planetary Model

A model of the solar system, the world's largest scale model, stretches 40 miles between Houlton and Presque Isle. The Houlton Information Center, just north of I-95 on U.S. 1, houses the planet Pluto and has a brochure about the model. The model of the Sun is in Folsom Hall at the University of Maine at Presque Isle. The actual distance between the Earth and the Sun is 93 million miles, known as an astronomical unit, and is represented by 1 mile in the model.

Carys Mills Formation

The Carys Mills formation lies in an irregular belt along the eastern border of Aroostook County, from Houlton almost to Van Buren. It was deposited during Ordovician and Silurian time, and consists of limy slate. The original sediments were mud and limy mud. Weathered exposures of the Carys Mills formation have indentations over the less resistant limestone beds, which weather to a light tan, while the ordinary mudstone remains black. The grooves and color bands inspire people to call it the ribbon rock, or the ribbon lime.

Some of the best exposures of the Carys Mills formation are at the type locality, about 2 miles south of Houlton on U.S. 2A, the Military

Ribbon lime of the Carys Mills formation in Monticello.

Road. The bedding in these exposures on the south side of the road and of the Meduxnekeag River has a nearly vertical dip. The layers trend parallel to the road.

Littleton Esker

U.S. 1 follows the west side of the Littleton esker for a number of miles. The esker is a winding ridge about 60 feet high. It began as the gravel bed of a meltwater stream that flowed in a glacier tunnel during the last ice age. As the ice melted, the tunnel sediments became winding ridges of sand and gravel.

For some reason, eskers are much more common in central and southern Maine than in Aroostook County. Some geologists think that may reflect the greater abundance of granite and sandstone in central and southern Maine. Those rock types more easily break down into sand and gravel than do the slaty sedimentary rocks of northern Maine.

Mars Hill

The town of Mars Hill is named for Mars Hill, the highest elevation in the eastern part of The County, 1,660 feet. It rises sharply above the Aroostook plain, which has an elevation of about 500 feet. The resistant Mars Hill conglomerate, which was deposited in Ordovician and Silurian time, holds up the top of the hills. The less resistant limestone and shale of the Carys Mills formation underlies the surrounding plain.

Prestile Stream and the Beet Pollution Incident

In the early 1960s, some farmers in Aroostook County began growing sugar beets when a large processing plant was built in Easton, near the head of Prestile Stream. The stream flowed through Mars Hill and Blaine and east into Canada.

The processing plant dumped wastes from sugar production into Prestile Stream, which had been known for its brook trout. The fish soon died. And the stream stank. Complaints fell on deaf ears in Augusta, perhaps because the state had helped fund the plant. Nothing was done. Then, in 1967, Prestile Stream flooded low ground in Blaine and Mars Hill with stinking purple water. The Canadians had built an earthen dam on the border to stop the flow of polluted water into New Brunswick, and they protected it with an armed guard until state officials agreed to abate the pollution.

In any case, sugar beets are a poor crop for Aroostook County. The growing season is too short for sugar beets, and their roots penetrate so deeply into the soil that mechanical harvesting is nearly impossible. Many people in The County had invested in the sugar beet project, and they lost heavily when it failed.

Aroostook State Park

A few miles south of Presque Isle, a road leads about 1 mile to Aroostook State Park. Quaggy Joe, a small peak that rises above Echo Lake, consists of the Hedgehog formation, a pale volcanic rock that belongs to the Dockendorf group of formations. These volcanic rocks are the northeastern end of the Acadian volcanic chain that also includes the Traveler volcanic rocks in Baxter State Park and the Kineo flint near Moosehead Lake.

Mapleton Formation

Outcrops of the Mapleton formation appear a few miles west of Presque Isle. The formation contains sandstone and conglomerate that were laid down in river floodplains during Devonian time. The outcrop area forms an almost perfect circle.

The Mapleton formation contains abundant fossils of land plants that grew during middle Devonian time, about 380 million years ago. They open a window on the early stages of the evolution of land plants. A buried erosion surface separates the Mapleton formation from the

Conglomerate and sandstone beds of the Mapleton formation in Mapleton.

sedimentary and volcanic rocks beneath it. They were folded during the Acadian mountain-building event, then eroded, before the Mapleton formation was laid down. That helps date the Acadian mountain-building event and establishes the Mapleton formation as the oldest sedimentary rock in this area that escaped its effects.

Three other Devonian sedimentary rocks in Maine contain fossils of Devonian land plants: The Trout Valley formation of the northern portion of Baxter State Park contains early Devonian land plants. It also overlies volcanic rocks erupted during the Acadian mountain-building event. The Fish River Lake formation between Fort Kent and Ashland also contains land plant fossils, as well as brachiopods, which lived in seawater. Perhaps the land plants washed out to sea. Between Calais and Eastport in eastern Maine, fossils of late Devonian land plants occur in the Perry formation, a conglomerate and sandstone that was deposited on an erosion surface developed on Devonian granite and Silurian basalt.

Haystack Mountain and the Winterville Formation

A few miles west of Mapleton on Maine 163, a sharply peaked hill, Haystack Mountain, rises from a long ridge. Both the mountain and the ridge are made of volcanic rocks assigned to the Winterville formation. They erupted during Ordovician time. The conical shape of Haystack Mountain is an accident of erosion that has nothing to do with volcanoes.

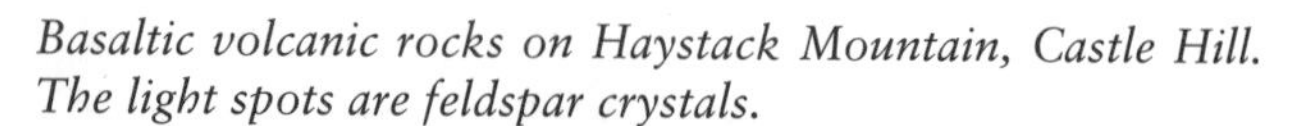

Basaltic volcanic rocks on Haystack Mountain, Castle Hill. The light spots are feldspar crystals.

Rocks exposed near Haystack Mountain range from basalt to rhyolite in composition. They almost certainly erupted from a chain of volcanoes similar to those now active in the Aleutian Islands. That chain lay parallel to a deep oceanic trench that was swallowing the floor of the Iapetus Ocean during Ordovician time. The chain extended from Newfoundland to the southern Appalachians. Volcanic rocks of the same age in northern England and Wales are probably part of the same volcanic chain. Opening of the North Atlantic Ocean separated them from their counterparts in north America.

New Sweden Formation

U.S. 1 crosses the New Sweden formation from the Little Madawaska River near the settlement of Acadia to Connor Township. Its outcrop area also extends southwest to near Mapleton, west of Presque Isle.

The New Sweden formation consists of sedimentary rocks ranging from conglomerate to slate and includes a good deal of calcareous slates similar to those in the Carys Mills formation. They were deposited in seawater during early Silurian time.

Although rocks of the New Sweden formation have about the same lime content as those in the Carys Mills formation, the soils differ in their potato-growing qualities. Soils on the New Sweden formation

Limy shale and sandstone of the New Sweden formation in Connor Township. The light gashes are filled with calcite.

are stony and have a high clay content that makes them hold water and prevents their draining after a rain. And the New Sweden formation erodes into a more rugged landscape than does the Carys Mills formation, making mechanical potato harvesting difficult.

Watch north of Caribou for the dramatic change in land use at the contact between the New Sweden and Carys Mills formations, near the Little Madawaska River. The land changes northward from nearly complete cultivation to forest cover and small gardens near homes. However, soils developed on the New Sweden formation do support agriculture in some upland areas.

Manganese Deposits

An ore deposit is a rock that can be mined and sold at a profit. The richest manganese deposits in Aroostook County are not now ores, because they are not now profitable, although many geologists think of them as ores. Deposits of manganese exist in at least three areas.

Some of the red and green slates of the New Sweden formation contain manganese oxides and carbonates. Similar minerals exist in the Maple Mountain formation, west of Bridgewater. The richest manganese deposits are near Dudley Road in the township of Castle Hill, west of Mapleton; they are in the New Sweden formation. Other deposits are on Maple Mountain and Hovey Mountain, west of Bridgewater.

No American mines produce much manganese; we import almost our entire supply for use in steelmaking. The Aroostook County deposits contain reserves of about 340 million tons of rock that runs about 9 percent manganese and 20 percent iron—or about 30 million tons of manganese and almost 70 million tons of iron. These are the largest manganese deposits in the country. So far, no one has found a way to mine them profitably, because of the difficulty in removing the manganese from the rock.

Van Buren and the St. John Valley

Van Buren was named for the eighth president of the United States, who was in office during the Aroostook War. Most of the people who live along the St. John Valley between Van Buren and Fort Kent are descendants of Acadian loggers and farmers who arrived in 1791. Once much persecuted, they have lived in peace on the St. John River for two hundred years.

The St. John River is one of the larger streams in the United States, measured in the volume of its flow. Its watershed receives ample rainfall in a cool climate that minimizes losses to evaporation and plants.

Madawaska Lake Formation

North of Van Buren, the Bangor and Aroostook Railroad crosses U.S. 1 near the settlement of Parent. Bedrock between Parent and Grand Isle is the Madawaska Lake formation, an Ordovician mudstone. To the southeast, the Madawaska Lake formation gives way to Ordovician volcanic rocks of the Winterville formation. These Ordovician rocks are exposed in a fold belt named the Munsungun-Winterville anticline. It trends northeast.

Grand Isle

Grand Isle is named for a large gravel bar in the middle of the St. John River. The bar is now about 5 miles downstream of the town named for it, probably because bars slowly migrate down the river. Floods erode gravel from the upstream ends of the bars and deposit it on the downstream end.

The Seboomook formation is exposed near the rest area on U.S. 1, in the northern part of Grand Isle. Strong sets of intersecting fractures break the sandy rock into rods about a foot long. Geologists call it pencil cleavage.

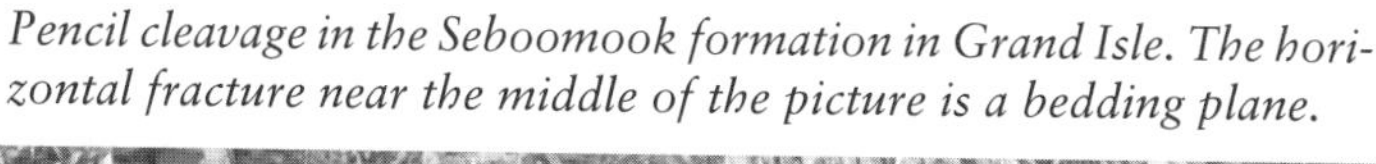

Pencil cleavage in the Seboomook formation in Grand Isle. The horizontal fracture near the middle of the picture is a bedding plane.

Madawaska

Madawaska is the northernmost town in Maine, the largest along the St. John River. From the Acadian French settlement at St. David to well above the junction of the St. John and the Allagash Rivers, the bedrock is the Seboomook formation, a great stack of slates and sandstones originally deposited during Devonian time.

The numerous gravel bars in this stretch of the St. John River are mined each year for sand and gravel. The annual spring floods replace the gravel. It is a renewable ore body, one of the few.

Glacial Lake Madawaska

A lake flooded the valley of the St. John River during the later stages of the last ice age. The type of dam remains unclear: perhaps it was the ice flowing from the New Brunswick highlands, or possibly a large moraine at Grand Falls, New Brunswick. Whatever, it backed up the St. John River at least as far as St. Francis. That was Glacial Lake Madawaska. Glacial outwash sediments deposited in this lake fill low areas along U.S. 1. Numerous gravel pits expose them.

St. Daniel and Depot Mountain Formation

The St. Daniel formation is a sequence of mudstones deposited during Cambrian and Ordovician time. The Depot Mountain formation includes seafloor basalt, old oceanic crust, and sedimentary rocks deposited during Ordovician time.

These are a small sliver, a northern extension, of the Cambrian and Ordovician rocks that appear in the Green Mountains of Vermont, the Berkshire Mountains of Massachusetts, and the Blue Ridge Mountains of the central and southern Appalachians. These rocks were laid down on the floor of the Iapetus Ocean, then jammed against the Grenville metamorphic rocks, the old Precambrian basement of eastern North America, during the Taconic mountain-building event.

These rocks are hard to see, hardly roadside geology. You can see some of them along a private wood road that meets Maine 161 at Dickey, or by driving north from Fort Kent to New Brunswick and then following Québec 289. This road leads to the northernmost settlement in Maine, Estcourt Station.

Maine 11
Sherman—Fort Kent
110 miles

The Aroostook Road connected Molunkus, in the southwestern part of The County, to Ashland; the Fish River Road connected Ashland to Fort Kent. Both were begun in 1831 to reach woods camps in the north and to assist American efforts to hold the lands claimed by Britain. Maine 11 more or less follows the portion of the Aroostook Road between Sherman and Ashland. Another road between Molunkus and Sherman is still known as the Aroostook Road and is certainly a remnant of the earlier one.

Most of Maine 11 crosses the Seboomook formation, interbedded sandstone and shale that were deposited during Devonian time. It also crosses two belts of older rocks exposed in the Munsungun-Winterville anticline and the Lunksoos-Weeksboro anticline.

The road crosses the Winterville fold twice, passing volcanic rocks of the Winterville formation, which erupted during Ordovician time. Ordovician volcanic rocks are also exposed in the Weeksboro structure, as is gabbro that may represent the magmatic source of these volcanic rocks. Other rocks also present, but not well exposed along Maine 11, include the Grand Pitch formation, a Cambrian mudstone and sandstone. About 20 miles of the road near Sherman cross the Allsbury formation, a brown mudstone deposited during Silurian time.

Copper Ore in Aroostook and Piscataquis Counties

Two ore bodies exist within about 15 miles of Maine 11. Both contain primarily copper, along with lesser amounts of gold, silver, and zinc. Both are in volcanic rocks that erupted during Ordovician time, during the Taconic mountain-building event. The Bald Mountain ore body is about 15 miles west and a bit north of Ashland. The Mt. Chase ore body is on the mountain of the same name, about 10 miles northwest of Patten.

Allsbury Formation

The roadcuts at the interchange between I-95 and Maine 11 reveal excellent exposures of the Allsbury formation of Silurian age. The rocks have nearly vertically dipping bedding and cleavage, although you can see some small folds. The rock owes its light brown color to the rather rare mineral siderite, a carbonate of iron. It is supposed to

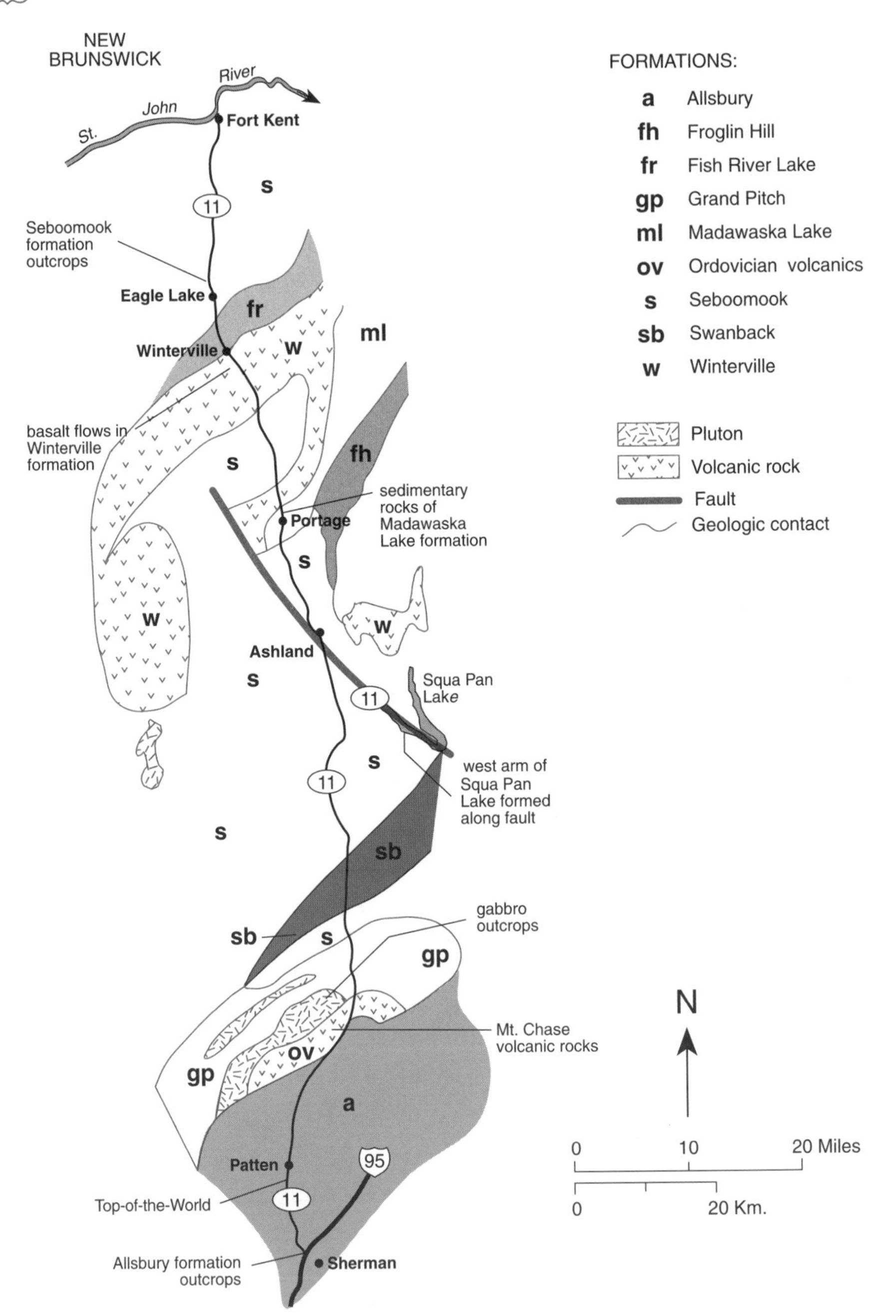

Geologic features along Maine 11, from Sherman to Fort Kent.

contain fossil graptolites, which commonly appear as black tracery on the bedding surfaces, but I have not found any.

Top-of-the-World

About halfway between Sherman Station and Patten is Ash Hill, known locally as Top-of-the-World. This is the highest hill around, and it is completely cleared, so you can see a full panorama of the surrounding territory. To the west are Katahdin and Traveler, eroded in granite and volcanic rhyolite, respectively. To the north are the conical peaks of the Mt. Chase volcanic rocks, of Ordovician age, the highest being Mt. Chase itself with its copper ore. The flat plains formed by the Carys Mills formation in eastern Aroostook County are obvious to the northeast. To the east are the low Oakfield Hills eroded in hornfels of the Allsbury formation, in the baked zones around small granite plutons. To the southwest, and visible on a clear day, are higher hornfels mountains in the baked zone of the Carrabassett formation around the Moxie pluton. Hornfels mountains continue southwest through Greenville, to Sugarloaf Mountain and the Rangeley Lakes region, some of the most continuously elevated land in Maine.

Mt. Chase Volcanic Rocks

Along the 5 miles north of Patten, Mt. Chase and its companion peaks loom west of the road. All are made of volcanic rocks that erupted during Ordovician time, during the Taconic mountain-building event. Maine 11 crosses this belt of volcanic rocks near the rest area near Halls Corner.

Watch just south of Knowles Corner for dark exposures of gabbro and diorite, the plutonic versions of basalt and andesite, respectively. These coarsely crystalline rocks may have crystallized in the magma chambers from which the Mt. Chase volcanic rocks erupted. The large volumes of magma crystallized deep below the surface into the larger mineral grains that generally distinguish plutonic from volcanic rocks.

Squa Pan Lake

Squa Pan Lake has a prominent **V** shape, with its apex pointing to the southeast. The basin of the narrower western limb is eroded along a fault in the Seboomook formation and is parallel with the movement of the last ice sheet. It is about 8 miles long. The eastern limb of the lake is 10 miles long and is formed in Devonian-aged shales. The lake lies mostly in Squa Pan Township southwest of Presque Isle. Its name is said to mean "bear's den" in the Algonquin language.

Winterville Formation

Bedrock along the 5 miles north of Portage is the Winterville formation, volcanic rocks that erupted during Ordovician time, and some sedimentary rocks of the same age. The volcanic rocks are essentially the same as those on Mt. Chase to the south. They erupted from a

Sedimentary rocks of the Winterville formation in Portage.

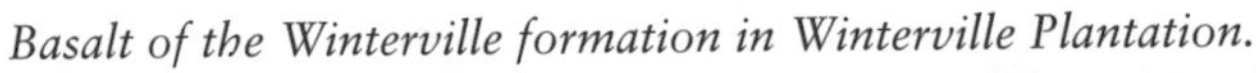
Basalt of the Winterville formation in Winterville Plantation.

chain of volcanoes that grew above the sinking slab of oceanic crust during the Taconic mountain-building event.

Farther north the road crosses a belt of the Seboomook formation; then, about 7 miles north of Portage, near a roadside rest area, it again crosses onto the Winterville volcanic rocks. In this area, the outcrops of basalt show small, but well-developed, pillow structures. They are good evidence that the flows erupted underwater.

Fish River Lake Formation

The Fish River Lake formation is a calcareous mudstone of Silurian and Devonian age that underlies the area between Winterville and Eagle Lake. It is poorly exposed along Maine 11. It is noted for its unusual fossil content: Both marine shellfish and land plants occur in it. The land plants must have grown on land raised above the ocean, fragments of which washed out to sea.

Between Eagle Lake and Fort Kent, a distance of about 12 miles, the route crosses the Devonian Seboomook formation. Watch for the excellent exposure at the top of the long hill south of Fort Kent. A roadside rest area is east of the road, the outcrop west. The rest area provides a fine view of the Fish River valley. The rocks are only moderately deformed, with the bedding still nearly horizontal in places.

Seboomook formation in Eagle Lake. The vertical planes are cleavage formed during folding. The light inclined lines are bedding planes.

Glossary

amphibole. A generally dark silicate mineral containing iron, magnesium, or calcium that occurs in granite and in some other igneous and metamorphic rocks.

andalusite. An aluminum silicate mineral that occurs in pelitic rocks that have been baked by a nearby intrusion. It forms stubby pink crystals.

andesite. A gray volcanic rock intermediate between basalt and rhyolite in composition. It forms mainly in volcanic chains near oceanic trenches.

anticline. A fold in which the layers bend up to make an arch.

arete. A narrow ridge between two glacially gouged valleys.

basalt. A dark volcanic rock. The lava is very hot, and fluid erupts quietly.

basement rocks. Ancient, highly metamorphosed rocks that underlie most other rocks in Maine. Basement rocks are exposed north of Stratton in the Chain Lakes massif.

beach cusps. Regularly spaced spurs on a beach.

bedrock. Rock that is part of the solid earth. Commonly known in Maine as ledge.

biotite. Black mica. It is a minor mineral in some granites and schists.

calcite. A mineral composed of calcium carbonate, the major component of limestone and marble.

cirque. A steep-sided basin, open at one side, carved by mountain valley glaciers. Mt. Katahdin has the largest number of cirques of any mountain in Maine.

conglomerate. Gravel cemented into solid rock.

DeGeer moraine. Rounded, short ridges of till, sand, and gravel and glacial marine mud deposited by glaciers lodged on the seafloor in several hundred feet of water.

delta. An accumulation of sand and gravel deposited in lakes or in the ocean. The flat top of the deposit is slightly above the water in which it formed.

dike. A sheet of igneous rock injected into a fracture.

diorite. A dark igneous rock intermediate in composition between granite and gabbro.

dunite. An igneous rock composed largely of olivine.

erratic. A boulder that differs from the underlying bedrock, and was therefore carried to its position.

esker. A winding ridge of sand and gravel deposited from streams flowing in tunnels beneath melting glaciers.

fault. A fracture in rocks along which the rocks on either side have moved.

feldspar. A group of minerals containing calcium, sodium, or potassium combined with aluminum silicate. They occur in igneous and some metamorphic rocks. Feldspars are the most abundant mineral in the earth's crust.

floodplain. A flat, river valley floor.

formation. A body of rock that can be recognized over a wide area.

fossil. The remains or an impression of a plant or animal.

gabbro. An intrusive igneous rock composed about half and half of pyroxene and calcic feldspar. Basalt is its volcanic equivalent.

garnet. A group of aluminum silicate minerals that also contains either iron, manganese, calcium, or chromium and forms 12-sided crystals. It occurs in granite, in some metamorphic rocks, and in stream and beach sands.

glacial till. A chaotic mixture of rocks, sand, and mud plastered on the ground by glaciers. Forms moraines and drumlins.

gneiss. A high-grade metamorphic rock in which light- and dark-colored minerals are segregated into wavy bands.

granite. An intrusive igneous rock composed largely of feldspar and quartz. It is the most common igneous rock on continents.

hornfels. A mudstone or shale baked in the heat of an igneous intrusion. The highest mountains in Maine, except Katahdin, are made of this tough, resistant rock.

hot spot. A locus of volcanic activity not related to a plate boundary.

igneous. A rock crystallized from molten magma, either on or beneath the surface.

lava. Molten rock that erupts to become volcanic rock.

lepidolite. A pink mica that contains lithium. Occurs in some pegmatites.

limestone. Sedimentary rock composed of calcium carbonate.

limy rock. Sedimentary rock containing some calcium carbonate.

magma. Molten rock beneath the surface.

mantle. The portion of the earth below the crust and above the core. The largest part of the earth.

marble. Metamorphosed limestone, made of larger calcite crystals than the original limestone.

meander. The smooth, sinuous course of a river flowing on a floodplain. The Sandy River and the Saco River have excellent meanders.

meltwater. Melted glacier ice that carries sand, gravel, and mud and deposits them as glacial outwash.

metamorphic. A rock recrystallized under high temperature, and in most cases high pressure.

mica. A family of silicate minerals that split into thin sheets.

moraine. Any landform made of glacial till. Many are ridges deposited at the side or front of a glacier.

oceanic crust. The crust of the earth that underlies ocean basins, composed mostly of basalt.

oceanic ridge. A ridge that underlies all ocean basins. Basalt magma erupts within it to make new oceanic crust as the opposite sides pull apart.

oceanic trench. A deep trough that develops where the ocean floor bends to start sinking into the mantle. The greatest oceanic depths occur in trenches.

olivine. A generally green iron and magnesium silicate mineral that occurs mainly in gabbro and peridotite.

ophiolite. Oceanic crust on land, generally as part of an accumulation of rocks swept off the sinking ocean floor and into an oceanic trench.

orthoclase. A pale and generally pinkish feldspar that contains potassium, a major component of granite.

outcrop. An exposure of bedrock.

outwash. Sand and gravel dumped from glacial meltwater.

pegmatite. A very coarse-grained granite.

phyllite. A mudstone or shale metamorphosed to a stage between slate and mica schist.

pillow basalt. Basalt that erupts underwater to make rounded lumps that suggest pillows. Most pillow basalt erupts in oceanic rifts.

placer. A segregation of heavy minerals in stream or beach sand. A few placers contain gold. The best placer in Maine is on the Swift River in Byron.

plagioclase. A family of silicate minerals that contain calcium and sodium. They occur in many types of igneous and metamorphic rocks.

pluton. A body of intrusive igneous rock. Most plutons are granite.

pyroxene. A generally dark group of silicate minerals that contain iron, magnesium, and calcium. Pyroxenes are a major component of basalt and gabbro, and occur in some other igneous and metamorphic rocks.

quartz. A mineral composed of silicon and oxygen that occurs in many kinds of rocks. Quartz—the commonest mineral on continents—comes in a wide variety of forms.

radiocarbon dating. A method of measuring the age of defunct plant or animal tissues by analyzing their residual content of carbon-14. Used widely in the study of deposits of the last ice age.

rhyolite. A volcanic rock with the same chemical and mineral composition as granite.

ribbon lime. A common rock in central Maine consisting of thin layers of limestone alternating with sandstone or mudstone.

sandstone. Sand solidified into rock.

schist. A generally sparkly metamorphic rock in which flat minerals like mica are arranged in sheets.

silicate. A mineral that contains silica.

sillimanite. Aluminum silicate mineral formed during extreme metamorphism of shale or mudstone.

slate. Mildly metamorphosed shale or mudstone that tends to split easily into flat sheets.

slaty cleavage. Parallel planes along which slate splits.

spodumene. A pyroxene that contains lithium. It occurs in pegmatites with lepidolite mica. Pink kunzite and green hiddenite spodumene are sometimes cut into gems.

syenite. An intrusive igneous rock in which all the silica is consumed in making feldspar and amphibole, leaving none to make quartz.

syncline. A fold in which the layers of rock bend downward into a trough.

terrane. A body of rocks that may be quite different, but share a similar origin and history.

tourmaline. A complex silicate mineral that contains a large variety of elements. It occurs mainly in pegmatites. Pink, green, and black tourmaline are common in Maine.

tuff. A rock composed mainly of volcanic ash.

unconformity. A break in the sedimentary record. Generally some layers are eroded and the erosion surface is buried under younger beds.

unicite. A pink and green granite formed by the alteration of feldspar. Much of the Attean pluton near Jackman is composed of unicite.

Additional Reading

Berry, A. W., Jr., ed. 1989. *Guidebook to field trips in southern and west-central Maine.* Farmington, Maine: 81st Annual Meeting of the New England Intercollegiate Geological Conference.

Bradley, Dwight C. 1983. Tectonics of the Acadian orogeny in New England and adjacent Canada. *Journal of Geology* 91:381-400.

Cheney, J. T., and C. Hepburn, eds. 1993. *Field trip guidebook for the Northeastern United States,* 2 vols. Boston: Geological Society of America Annual Meeting and 85th Annual Meeting of the New England Intercollegiate Geological Conference. Ten of the field trips cover Maine geology.

Drake, A. A., and others. 1989. The Taconic Orogen. In *The Geology of North America, Vol. F-2, The Appalachian-Ouachita orogen in the United States,* eds. R. D. Hatcher, Jr. and others. 123. Boulder, Colorado: Geological Society of America.

Gaer, Joseph, ed. 1937. *Maine, A Guide "Down East."* Boston: Houghton Mifflin Company.

Hanson, L. S., and D. W. Caldwell, eds. 1994. *Guidebook to field trips in northern and central Maine.* Millinocket, Maine: 86th Annual Meeting of the New England Intercollegiate Geological Conference.

Hussey, A. M., ed. 1995. *Guidebook to field trips in Southern Maine and adjacent New Hampshire.* Brunswick, Maine: 87th Annual Meeting of the New England Intercollegiate Geological Conference.

Ludman, A., ed. 1992. *Guidebook to field trips in eastern coastal Maine and adjacent New Brunswick.* Princeton, Maine: 84th Annual Meeting of the New England Intercollegiate Geological Conference.

Newburg, D., ed. 1986. *Guidebook for field trips in southwestern Maine*. Lewiston, Maine: 78th Annual Meeting of the New England Intercollegiate Geological Conference.

Osberg, P., A. M. Hussey, and G. M. Boone, eds. 1985. *Bedrock geologic map of Maine*. Augusta: Maine Geological Survey. Scale 1:500,000.

Roy, D. C., ed. 1987. *Centennial Field Guide, Vol. 5*. Geological Society of America. Nine of the field trips cover Maine geology.

Smith, G. W. 1980. *End moraines and glaciofluvial deposits, Cumberland and York Counties, Maine*. Augusta: Maine Geological Survey, map.

Thompson, W. B., D. L. Joyner, R. G. Woodman, and V. T. King. 1998. *A collector's guide to Maine mineral localities*. 3d ed. Augusta: Maine Geological Survey Bulletin 41.

Thompson, W. B., and H. W. Borns, eds. 1985. *Surficial geologic map of Maine*. Augusta: Maine Geological Survey. Scale 1:500,000.

Tucker, R. D., and R. G. Marvinney, eds. 1989. *Studies in Maine Geology*. Augusta: Maine Geological Survey. Six volumes on all aspects of the geology of Maine.

Williams, Harold. 1978. *Tectonic lithofacies map of the Appalachian orogen*. St. John's: Memorial University of Newfoundland.

Index

About the Author

D. W. Caldwell was raised in rural Maine where, at the age of twelve, he decided to become a geologist when he earned a Boy Scout geology merit badge. He obtained a Ph.D. in geology from Harvard University and worked for both the United States and the Maine Geological Surveys. Caldwell was a consultant in construction materials, in bottled water sources, and in water pollution studies and served as an expert witness in cases dealing with these issues. He was a professor of geology at Boston University for more than thirty years. He passed away in 2006.